# The Business of Densification

Gabriela Debrunner

# The Business of Densification

## Governing Land for Social Sustainability in Housing

palgrave
macmillan

Gabriela Debrunner
Institute for Spatial and Landscape
Development IRL
ETH Zürich
Zürich, Switzerland

ISBN 978-3-031-49013-2 ISBN 978-3-031-49014-9 (eBook)
https://doi.org/10.1007/978-3-031-49014-9

The open access publication of this book has been published with the support of the Swiss National Science Foundation



Cover illustration: Contributor: alimdi.net/Alamy Stock Photo

This Palgrave Macmillan imprint is published by the registered company Springer Nature Switzerland AG
The registered company address is: Gewerbestrasse 11, 6330 Cham, Switzerland

Paper in this product is recyclable.

*The large question is not whether abandonment can be avoided, gentrification controlled, displacement eliminated, or even how these things can be done, but rather whether there is the desire to do them. That is a question that can only be answered in the political arena.*

*Marcuse, 1986:175*

# Promotion Committee

**Supervisor**
Prof. Dr. Jean-David Gerber
Political Urbanism and Sustainable Spatial Development &
Centre for Regional Economic Development
Institute of Geography
Faculty of Science
University of Bern

**External referee**
Prof. Dr. Barbara Schönig
Urban Planning
Faculty of Architecture and Urbanism
Bauhaus-University Weimar

**Chairwoman**
Prof. Dr. Heike Mayer
Economic Geography & Centre for Regional Economic Development
Institute of Geography
Faculty of Science
University of Bern

# Preface and Acknowledgments

The housing question has always concerned me. Why do some people live at certain locations while others do or cannot? Why and how do such social differences emerge? Who decides what, why, and how? My human geography studies at the Institute of Geography at the University of Bern taught me a lot about the political decision-making processes behind housing dynamics. However, even after completing my master's studies, I was still looking for concrete answers.

At this point, I would like to thank my supervisor and mentor Prof. Dr. Jean-David Gerber, who significantly influenced my way of thought with his view of the world and his understanding of the mechanisms at play affecting sustainable housing use. His comments on drafts provided me with the opportunity to pursue my ideas in research and teaching, which strongly shaped my identity as a critical human geographer. I am very grateful for his trust, support, and encouragement throughout my studies in Bern. Writing my dissertation at a chair that promotes new theoretical approaches while being very socially critical towards ongoing environmental challenges was one of the best decisions I have ever made.

I would also like to thank Prof. Dr. Barbara Schönig for her external review of my thesis. I have been following her work for many years and highly appreciate the fact that a scholar of her experience and expertise at the interface between land use policy, spatial planning, and affordable housing provision reviewed my work. Thank you very much.

My thanks also go to Prof. Dr. Heike Mayer for her internal chairwomanship. She has accompanied me on my way at the Institute for many years. Her economic geography perspective and feedback on planning challenges as well as her knowledge of the Swiss political and spatial planning system have always helped me in improving my work. Many thanks for your support.

My gratitude also goes to my team at the Institute of Geography—Andreas Hengstermann, Ivo Balmer, Kristina Lanz, Elke Kellner, Christoph Oberlack, Timothy Adams, James Natia Adam, Deniz Ay, Josje Bouwmeester, Jessica Verheij, Sybille Vogel, Andrea Weber, Aron Affolter, Nico Rast, Stephan Kägi, Marlen Stöckli, and Louis Zwyssig—for the many nice lunches and discussions we had together. A big thank goes to my former colleague Andreas Hengstermann, who took me under his wings the moment I started my thesis and showed me how life as a Ph.D. candidate works at the Institute. My gratitude also goes to my peers from the Lower and Upper Midstaff Faculty—Andrea Winiger, Rahel Meili, Reto Bürgin, Maaret Jokela, Elisabeth Miliz, Jeannine Wintzer and Alexander Vorbrugg—who always supported and accompanied me.

Moreover, my thanks go to my dear friends from the Planning, Law, and Property Rights (PLPR) community—Thomas Hartmann, Arend Jonkman, Rick Meijer, Mathias Jehling, Sebastian Dembski, Cygal Pellach, Linda McElduff and Arthur Schindelegger—who supported me on my way and for the many interesting conversations and academic assistance. The exchange of perspectives has always helped me to develop. In addition, many thanks to Nina Steensen, who provided me with the opportunity to perform a research stay at the International Federation for Housing and Planning (IFHP) in Copenhagen in order to exchange ideas and collaborate. Our discussions together with Anton Øsgårds and Morton Nielsen during my time in Copenhagen were highly stimulating and helped me to further develop my mind as a researcher.

Finally, I am deeply grateful to my family, my sisters, my friends, and my partner Witali! Especially, I would like to thank my parents who have made this path possible for me and who always supported my decisions. I cannot thank them enough for the opportunity to complete my education and for the fact that they mastered it all with three children. My greatest thanks are also designated to Witali for all his patience and support on countless occasions and in periods in which I doubted my process. Writing a dissertation is ultimately a very great, but also lonely experience, and I would like to thank him for always accompanying and supporting me.

Thank you all!

Heard Island and McDonald Islands

Naveen Dass

# Competing Interests

The author has no conflicts of interest to declare that are relevant to the content of this book. The publishing process of this book was funded by the Swiss National Science Foundation (SNSF) via a four-year project titled "Densifying Switzerland" (2021–2025) (Grant number: 200499), and the Future Cities Laboratory (FCL) Global of ETH Zürich.

# Ethics Approval

The studies of this book were performed in line with the principles of the Declaration of Helsinki.

Approval was granted by the Ethics Committee of the University of Bern.

# Contents


1 **Introduction** 1
- 1.1 *Push Towards Urban Densification in Cities Worldwide* 1
- 1.2 *Rise of the Compact City Model in International Policymaking* 1
- 1.3 *Densification Evokes Social Exclusion and Gentrification in Housing* 3
- 1.4 *Research Gap: Politicizing and Identifying Densification as a Governance Challenge* 5
- 1.5 *Research Objectives and Questions* 6
- *References* 8

2 **Governing by New Institutionalist Political Ecology** 13
- 2.1 Part I: *Theoretical Approach—Actors-Centered New Institutionalist Political Ecology* 13
  - 2.1.1 *From Classic to New Institutionalism: Historical Background of Institutional Thought* 14
  - 2.1.2 *From Hardin to Ostrom: Towards Political and Actor-Centered Neoinstitutionalism* 16
  - 2.1.3 *The Question of Power: Bringing a Political Ecology Perspective to Neoinstitutionalist Research* 19

2.2 Part II: *Analytical Framework—The Institutional Resource Regime (IRR) and Its Focus on Property Rights* 22

2.2.1 *The Institutional Regime: Two Sources of Formal Rules Regulating Resources Use* 23

2.2.2 *Public Policies* 25

2.2.3 *Property Rights—A Chronological Review* 27

2.2.4 *Conflict Relationship Between Public Policy and Property Rights* 32

2.2.5 *Actors' Use Strategies* 33

*References* 35

**3 The IRR Applied to Housing: Governing Densification for Socially Sustainable Housing Development** 43

3.1 *Housing Study Focus: Governing Densification for Socially Sustainable Housing Development* 43

3.2 *Governance of Densification for Sustainable Urban Housing Development* 43

3.3 *Housing Stock: Evaluating the Condition of a Resource* 45

3.4 *Housing Sustainability: With a Focus on Its Social Dimension (Dependent Variable)* 48

3.4.1 *Social Sustainability as an Analytical Framework of Reference* 49

3.4.2 *Social Sustainability in Housing* 54

3.4.3 *Social Sustainability in Housing as Emic Approach* 58

3.5 *Institutional Regime Regulating the Housing Stock (Independent Variable)* 60

3.5.1 *Housing Public Policies* 60

3.5.2 *Housing Policy Instruments* 61

3.6 *Actors' Housing Use Strategies (Intermediary Variable)* 62

3.6.1 *Active Municipal Land Policy for Sustainable Housing* 63

3.6.2 *The Power of Housing Property Owners as Titleholders* 65

3.6.3 *Tenants' Housing Resistance and Decommodification Strategies* 67

3.7 *Hypotheses: Governance Mechanisms Leading to Socially Sustainable Housing Outcomes in Densifying Cities* 70
*References* 73

**4 Selecting Switzerland** 87
4.1 *Study Design and Methodology: Densification and Urban Housing Development in Switzerland* 87
4.2 *Study Context: Challenges of Housing, Urban Land Scarcity, and Social Exclusion in Switzerland* 88
4.3 *Case Selection: Housing Under Densification Pressure in Four Swiss Cities* 91
4.3.1 *The Case of Zurich City* 94
4.3.2 *The Case of Basel-City* 98
4.3.3 *The Case of Köniz City* 99
4.3.4 *The Case of Kloten City* 101
4.4 *Data Collection Methods* 103
4.4.1 *Document and Statistical Data Analysis* 105
4.4.2 *Participant Observation* 106
4.4.3 *Household Surveys* 107
4.4.4 *Semi-Structured and Expert Interviews* 108
4.5 *Data Analysis Methods* 109
4.6 *Positionality, Reflexivity, and Validity of the Data* 110
*References* 111

**5 Investigating Switzerland** 117
5.1 *Investigating Switzerland: Learning from the Swiss Scarce Land Use Situation* 117
5.2 *Article 1: The Business of Densification—Distribution of Power, Wealth, and Inequality in Swiss Policymaking* 123
5.2.1 *Shift Towards Densification in Swiss Land Use Planning* 124
5.2.2 *Densification at the Interplay Between Public Policy and Property Rights* 126
5.2.3 *Housing from a Social Sustainability Perspective* 128
5.2.4 *Case Study of Densification Policies in Switzerland* 129
5.2.5 *Densification Policy Implementation and Social Sustainability in The Swiss Housing Sector* 131
5.2.6 *Towards a Business of Densification* 143

5.3 *Article 2: Planning for Social Sustainability: Mechanisms of Social Exclusion in Densification Through Large-Scale Redevelopment Projects in Swiss Cities* 146
5.3.1 *Implications of Densification on Social Sustainability in Housing* 147
5.3.2 *Planning for Social Sustainability in a Dense City* 149
5.3.3 *Comparative Case Study Analysis in Zurich Brunau and Basel Schorenweg* 152
5.3.4 *The Tensions Between Densification and Social Exclusion in Swiss Urban Policymaking* 157
5.3.5 *Explaining the Mechanisms at Play Leading to Residents' Social Exclusion in Zurich Brunaupark and Basel Schorenweg* 168
5.3.6 *The Eco-Business of Densification and the Neglection of Its Social Dimension* 170
5.4 *Article 3: Strategic Use of Land Policy Instruments for Affordable Housing—Coping with Social Challenges Under Scarce Land Conditions in Swiss Cities* 172
5.4.1 *The Role of Municipal Planning Authorities in Affordable Housing Provision* 173
5.4.2 *Affordable Housing Provision at the Interface Between Public Policy Intervention and Property Rights* 174
5.4.3 *Comparative Case Study Analysis of Four Swiss Cities* 177
5.4.4 *Strategic Use of Land Policy Instruments for Affordable Housing* 179
5.4.5 *Four Legal Intervention Ways to Generate Affordable Housing Outcomes* 198
5.4.6 *Towards an Active Municipal Land Policy* with *Property Rights* 202
5.5 *Article 4: The Commodification of Temporary Housing in Zurich* 204
5.5.1 *Temporary Housing in Densifying Cities* 205
5.5.2 *Development Stages of Temporary Housing* 206
5.5.3 *Temporary Housing in the City of Zurich* 211

5.5.4 *The Commodification of Temporary Housing in Zurich* 215
5.5.5 *Discussion: All in Favor of Commodified Temporary Housing?* 228
5.5.6 *Housing Is a Matter of Tenancy Law, Not Loaning Law!* 230
*References* 232

**6 Discussion of Results** 247
*6.1 Discussion of Key Results* 247
*6.2 Intermediate Summary* 257
*References* 259

**7 Final Conclusion: Governance Mechanisms for Socially Sustainable Urban Densification** 261
*7.1 Policy Recommendations: Towards Active Municipal Land Policy for Socially Inclusive Urban Housing in Densifying Cities* 264
*7.2 Theoretical and Methodological Limitations* 266
*7.3 Implications for Future Research: Densification, The IRR, and Beyond* 268
*References* 270

**Appendix** 273

**Index** 277

# About the Author

**Dr. Gabriela Debrunner,** Phil. Nat., is a postdoctoral researcher at the Institute for Spatial and Landscape Development (IRL) at ETH Zurich. She is associated at the Chair of Spatial Development and Urban Policy (SPUR) and has a leading role in the research projects "Densifying Switzerland" (2021–2025) (Swiss National Science Foundation SNSF funding) as well as "Dense and Green Cities" (Future Cities Lab FCL Global Program). Gabriela holds a Ph.D. in human geography and environmental social sciences with a focus on spatial planning and political urbanism from Bern University. She also has a Certificate of Advanced Studies (CAS) in Urban Management from the University of Zurich and a high school teaching diploma (gymnasium level) from the Bern University of Applied Sciences. In her research, Gabriela deals with the overarching question of how the city as socio-political space works: her main interests rely on densification as current challenges for planning and policymaking, gentrification, segregation, and social exclusion studies as well as active municipal land policy from urban governance and new institutional economy perspective. She thereby combines the three research fields of planning and housing studies with critical urban geography. During her research, Gabriela Debrunner spent several months at the University of Stockholm, Department of Human Geography, and in Copenhagen at the IFHP—International Federation for Housing and Planning. Before Gabriela joined ETH Zurich, she gained a wide range of experience in the private, public, and non-profit sectors. For her accomplishments, Gabriela Debrunner received the Faculty Price of Geosciences of the University of Bern for the best doctoral thesis in environmental sciences in 2021.

# Abbreviations

| | |
|---|---|
| AZMA | Association of Zurich Metropolitan Area |
| BTA | Brunaupark Tenants Association |
| CPR | Common-Pool-Resource |
| DIY | Do-It-Yourself |
| ETH | Swiss Federal Institute of Technology |
| EU | European Union |
| FEA | Federal Act on Energy |
| FHO | Federal Office for Housing |
| FHSA | Federal Housing Support Act |
| FOC | Federal Office for Culture |
| FOE | Federal Office for Energy |
| FOSD | Federal Office for Spatial Development |
| FOSI | Federal Office for Social Insurances |
| FSO | Federal Office for Statistics |
| FSPA | Federal Spatial Planning Act |
| FTA | Federal Tax Administration |
| IAD | Institutional Analysis and Development Framework |
| IFHP | International Federation for Housing and Planning |
| IRR | Institutional Resource Regime |
| IUCN | International Union for Conservation of Nature |
| LBC | Local Building Committee |
| MC | Municipal Constitution |
| NCHA | Federal Act for the Protection of Nature and Cultural Heritage |
| NGO | Non-Governmental Organization |
| NIE | New Institutional Economics |
| NIMBY | Not-In-My-Backyard |

| | |
|---|---|
| OC | Federal Obligations Code |
| PE | Political Ecology |
| PP | Public Policy |
| PPP | Public Private Partnership |
| PR | Property Right |
| SC | Swiss Constitution |
| UN | United Nations |
| WCED | World Commission on Environment and Development |

# List of Figures


Fig. 1.1 Schematic overview of the different elements and sections of this book 7

Fig. 2.1 Instruments of land policy (own Figure based on Balmer & Gerber, 2017; Gerber et al., 2018: 16) 27

Fig. 3.1 Governance of densification for sustainable urban housing development (overview): The IRR analytical framework applied to the study of housing stocks in dense city areas (Gerber et al., 2009, 2020; Knoepfel et al., 2007; Nicol & Knoepfel, 2008) 45

Fig. 3.2 The Institutional Resource Regime postulates a causal relationship (hypotheses 1–5) between three main variables: the condition of the housing- stock resource (dependent variable), actors (users) (intermediary variable), and institutions (independent variable) (Knoepfel et al., 2007; Gerber & Debrunner, 2022; Gerber et al., 2009, 2020: 157) 71

Fig. 4.1 Housing development in Switzerland since 1978 (Index based: 1978 = 100). While mortgage rates have constantly decreased since 1978, prices for land and rents havesteadily risen (Canton of Zurich, 2020; FOH, 2020; FSO, 2020a, 2020c; Wüest & Partner, 2020) 90

Fig. 4.2 Map of Switzerland and the cases conducted within the country—Zurich, Basel, Köniz, and Kloten (own figure) 93

Fig. 4.3 Densification areas within the four municipalities studied—Basel Schoren (top left), Kloten Southern district (top right), Köniz Nessleren (bottom left), and Zurich Brunau (bottom right) (own figure) 95
Fig. 4.4 Map of Brunaupark settlement located at the South-Western part of the City of Zurich (top), and two pictures of the inner-yard of settlement itself (below) (*Sources* GIS data system, Canton of Zurich, 2022; own photographs) 97
Fig. 4.5 Map of Brunaupark settlement located at the South-Western part of the City of Zurich (top), and two pictures of the inner-yard of settlement itself (below) (*Sources* GIS data system, Canton of Zurich, 2022; own photographs) 97
Fig. 4.6 Map of Basel Schorenweg settlement located at the North-East part of the City of Basel (top), and pictures of the inner-yard of settlement as well as renovated kitchens and bathrooms (below) (*Sources* Geodata Canton Basel-City 2021/2023; own photographs) 100
Fig. 4.7 Map of Basel Schorenweg settlement located at the North-East part of the City of Basel (top), and pictures of the inner-yard of settlement as well as renovated kitchens and bathrooms (below) (*Sources* Geodata Canton Basel-City 2021/2023; own photographs) 101
Fig. 4.8 Map of Köniz Nessleren settlement located at the Eastern part of the City of Köniz (top), and pictures of the former (left) and newly transformed and densified housing stock (right) (*Sources* Geodata Canton Bern 2021/2023; own photographs) 101
Fig. 4.9 Map of Köniz Nessleren settlement located at the Eastern part of the City of Köniz (top), and pictures of the former (left) and newly transformed and densified housing stock (right) (*Sources* Geodata Canton Bern 2021/2023; own photographs) 102
Fig. 4.10 Map of Kloten Southern district (*Source* GIS Browser Canton of Zurich, 2023) 104
Fig. 4.11 Photographs of Kloten Southern district settlement before (top) and after (below) demolishment and total replacement construction and densification (*Sources* Google Streetview; own photographs) 105
Fig. 5.1 Book structure: actors' triangle and user appropriation strategies at different geographical scales 118

| | | |
|---|---|---|
| Fig. 5.2 | Overview of the results of the household survey conducted in Zurich Brunaupark and Basel Schorenweg. Results show how tenants are affected by densification from a social sustainability perspective | 162 |
| Fig. 5.3 | Municipal land policy instruments for affordable housing in Swiss municipalities (analysis based on e.g. FOE & FOH, 2015; FOE, 2015, 2018; FOF et al., 2009; FOH 2012, 2014, 2014a, 2014b, 2016, 2016a, 2017, 2018b, 2019; FOSD, 2016; FOSI & FOH, 2015; Swiss Federal Council, 2014, 2016a, 2017) | 181 |
| Fig. 5.4 | The emergence of for-profit temporary housing in Zurich can be explained by the twofold process of institutionalization and commodification | 217 |

# List of Tables


| | | |
|---|---|---|
| Table 3.1 | Conceptualizations of social sustainability | 52 |
| Table 3.2 | Social sustainability indicators in housing. Analyzing social sustainability in densification processes takes place in close connection with ecologic and economic sustainability | 55 |
| Table 4.1 | Housing market characteristics in Zurich, Basel, Köniz and Kloten City (City of Kloten, 2020b; City of Köniz, 2020b; FOSD, 2017: 25; FSO, 2020b; Glattal Region, 2020; Nebel et al., 2017; Statistical Offices Zurich & Basel-City, 2019, 2020) | 92 |
| Table 5.1 | Overview of the four research articles composing this book | 121 |
| Table 5.2 | Formal federal institutions and policy instruments impacting social sustainability of housing under densification. The data analysis is based on an in-depth study of policy papers, legislations, and parliamentary debates (Flurį, 2017; FOC, 2018; FOE & FOH, 2015; FOE, 2018; FOF et al., 2009; FOH, 2012, 2014a, 2014b, 2014c, 2016a, 2016b, 2017; FOSD, 2016, 2018; FOSI & FOH, 2015; FTA, 2017; Hardegger, 2017; Swiss Federal Council, 2014, 2016a, 2016b, 2017, 2018; von Graffenried, 2014) | 137 |
| Table 5.3 | Housing market characteristics in Zurich and Basel-City (FOSD, 2017: 25; Statistical Offices Zurich and Basel-City, 2019, 2020) | 154 |

Table 5.4 Socio-economic profile and household types of tenants in Zurich Brunaupark & Basel Schorenweg 156
Table 5.5 Use of the land policy instruments for affordable housing provision in each city 187
Table 5.6 Steps, aims, and methods employed in this study 214
Table 5.7 Characteristics of the seven investigated temporary housing projects in Zurich in the year 2015 216
Table 5.8 Main dimensions of the regulatory framework influencing temporary housing practice in Zurich 218
Table 5.9 The two stages towards the commodification of temporary housing in Zurich 222
Table 5.10 Comparable example of a 4.5 room temporary housing apartment under letting and loaning (Wüest & Partner, 2017, and questionnaire with temporary users) 227

CHAPTER 1

# Introduction

## 1.1 Push Towards Urban Densification in Cities Worldwide

Providing shelter for a global population that is urbanizing quickly presents a critical human development challenge. By 2030, it is estimated that almost five billion of us will be living in cities, constituting around 60% of the world's population. Rapid urbanization is linked to numerous socio-environmental concerns such as concentrated use of energy and air pollution with significant impacts on human health, infrastructure, and economic prospects (UN, 2018). Facing these challenges has resulted in a myriad of solutions being proposed, and the list of possible urban forms within which dwellers can be housed sustainably is long and growing. High up on this list is the "compact city", which has been introduced as a promising way to meet the challenges of a growing urban population for a considerable time (e.g. Newman & Kenworthy, 1999).

## 1.2 Rise of the Compact City Model in International Policymaking

The "compact city" (in the US also termed "new urbanism" or "smart growth", "Stadt der kurzen Wege" in Germany) is a concept that evolved in the UK during the 1960s as parts of wider efforts to combat resource depletion (for discussion see e.g. Holden, 2004; Kahn, 2000). The


G. Debrunner, *The Business of Densification*,
https://doi.org/10.1007/978-3-031-49014-9_1

authors of these studies argue that compactness of the built environment—generally defined as an increase *in the density of units within city boundaries* (Boyko & Cooper, 2011: 47)—would slow down urban sprawl in order to limit settlement expansion and ensure sustainable urban growth.

More precisely, many advantages of the compact city model have been highlighted in the past few decades. They include, for instance, the conservation of the countryside (Elkin et al., 1991); the protection of environmentally vulnerable landscapes (Dieleman & Wegener, 2004); less need to travel by car, thus reducing fuel, energy, and air emissions (Ewing, 1997); the support for public transport, walking, and cycling modes of mobility (Squires, 2002); better access to services and facilities, along with more efficient utility and infrastructure provision (Frey, 1999); as well as increased potential for revitalization and regeneration of inner urban areas (Kahn, 2000). The compact city has become a physical response to many urban challenges, such as land consumption, energy and resource waste, accessibility, and air pollution. It has practically evolved as a synonym for "the sustainable city" (Neuman, 2005: 17).

Indeed, many international organizations, politicians, and urban practitioners have agreed with the benefits proposed and started to introduce "densification" as a legally binding policy objective (e.g. UN Declaration on Environment and Development 1992,[1] Principles 4 and 15). Densification (also termed "intensification" or "consolidation") is the process through which the compact city model is attained physically. A useful definition of the term can be found in Broitman and Koomen (2015: 32) who define densification as "a process leading to an increase in the number of households within existing municipal boundaries". The process creates an increase in exploitation or use density—defined as the *number of persons* per *square meter* (Boyko & Cooper, 2011: 47)—in order to reduce individuals' overuse of natural resources, such as land, water, or energy (Holman et al., 2015). Densification is thus widely assumed to play a decisive role in the sustainable transformation of settlements.

[1] In June 1992, at the Earth Summit in Rio de Janeiro more than 178 countries adopted Agenda 21, a comprehensive plan of action to build a global partnership for sustainable development.

## 1.3 Densification Evokes Social Exclusion and Gentrification in Housing

However, by the mid-1990s, multiple studies (e.g. Breheny, 1997; Cernea, 1993; Gordon & Richardson, 1997; Jenks et al., 1996) claimed that the process of implementing densification—next to its potential environmental benefits—would threaten the quality of life, particularly in regard to social aspects and the conditions of the poor (for discussion, e.g. Daneshpour & Shakibamanesh, 2011). The critique of the compact city concept focused on claims arguing that densification has undesirable social consequences. Those consequences include affordable housing shortage, residential displacement, and social exclusion as a direct consequence of (re)development and upgrading.

More precisely, densification has been accused of posing a threat to the very existence of *social sustainability in housing*, which focuses on various dimensions, such as social mixing, inclusion, residential stability, or neighborhood cohesion (see Section 3.3 for details). It can subsequently lead to unfair distribution of power and resources, freedom, access to decision-making, and general capacity-building (for discussion see e.g. Williams et al., 2000; Whitehead, 2003). Social sustainability in housing is generally defined as given if housing development "is compatible with harmonious evolution of civil society, fostering an environment conductive to the compatible cohabitation of culturally and socially diverse groups while at the same time encouraging social integration, with improvements in the quality of life for all segments of the population" (Polese & Stren, 2000: 1516) (for a detailed definition of the concept, Chapter 3.3).

While some approaches have pointed out the gentrification and displacement effect of densification on local residents and activities (Chiu, 2003; Lees, 2000), or the risk of weaker social ties in higher density environments (Freeman, 2001), others have examined the exacerbation of social exclusion of particular groups within local communities (Gosling, 2008) as well as the accumulation of residents' low skill jobs as results of displacement (Law, 2002). In addition, Williams et al. (2000) conducted research on the social sustainability of housing areas where development has been densified. Their often-quoted study concluded with claims that densification would result in a reduction of private space, smaller houses, and gardens, or no gardens at all. Moreover, more intensive traffic causes potential negative environmental impacts, such as air pollution, noise, and a generally poor environment for cyclists and pedestrians, as well as increase in potential "bad neighbor" effects, such as noise, disturbance, or litter (Williams, 2010).

In essence, the main argument against the compact city was and still is that densification leads to residents' social exclusion because there is an increase in housing prices and rents when real estate stocks are rebuilt and modernized. Although densification enables more apartments on the same parcel to be constructed, implementation tasks increasingly take place in the format of redevelopment of existing stocks e.g., through subdivision, extension to existing buildings, or total replacement construction since free inner-city urban green and brownfields are overbuilt already (Touati-Morel, 2015).

The result of such densification activities (e.g., renovation, subdivisions, replacements) within urban boundaries is that new housing is (re)built with higher densities—but also land prices and rents (Davidson & Lees, 2005). Newly built densified housing is thus often only accessible to certain—mostly high-income—groups of the population (Hackworth & Smith, 2001). Consequently, an increasing number of people worldwide is suffering from rising housing prices and rents as results of new-built densification (Aalbers & Christophers, 2014; Aalbers, 2017; Christophers, 2022). Affordable housing shortage and residents' social displacement have thus become severe societal problems in many cities globally (Rolnik, 2013; Wicki et al., 2022). Particularly lower income segments are forced to leave the city centers for cheaper suburban areas as they can no longer afford a dwelling in recently densified areas (Lees, 2008). Such social exclusion scenario, however, is considered highly unsustainable.

*Social exclusion* in turn is a process through which the composition of inhabitants changes, particularly due to the inflow of higher income groups and the (in)direct displacement of lower income groups, which in turn cause gentrification, social segregation, and social polarization (Lees, 2008: 2463). Hence, the result is that lower income residents living in these neighborhoods are forced to leave the center for cheaper suburban areas as they can no longer afford a dwelling in recently densified areas (Marcuse, 1985: 207). A vicious circle is created "in which the poor are continuously under pressure of displacement and the wealthy continuously seek to wall themselves within gentrified neighborhoods" (Marcuse, 1985: 196).

Dwellers deprived of housing in the center, and who are pushed to the margins of cities, struggle to find alternative housing in the city center. This is because they can no longer afford the rent after densification occurs. They are thus forced to move to the agglomeration areas, where rents are lower. This process may finally cause low-income groups to lack control over the most basic components of life—which are the places they call home (Slater, 2009: 307). Such a scenario is considered highly unsustainable (Jenks et al., 1996: 84). Hence, through densification, the solution to one problem (natural resource consumption) causes another problem (social exclusion) instead.

## 1.4 Research Gap: Politicizing and Identifying Densification as a Governance Challenge

In summary, an intense debate has been ongoing for more than forty years over the supposed environmental advantages of the compact city. That debate is also about the potential drawbacks of densification for different categories of stakeholders, particularly those of lower income. Whereas until the 1990s, development on greenfield outside city boundaries was largely promoted by policymakers and urban practitioners (for discussion, Filion, 2015), the land use conditions have changed under the compact city model. Land use interests cannot continue to be generously realized because unbuilt land is no longer available unlimitedly. Instead, implementing densification objectives implies that the needs and visions of different individuals and groups clash at the very same locations within the urban built environment.

The policy shift towards densification has made conflicting use interests more pronounced because stakeholders now must deal with each other in a context of scarce urban land. This implies that they must negotiate for their interests within the already built environment and within existing city boundaries. Implementing densification policy objectives becomes a complicated process because the objectives are embedded in a tight web of already existing, diverse, and contradictory rights, claims, and duties. What benefits one stakeholder potentially hurts another. A landlord's profits through (re)development or upgrading might come at a tenant's expense. High-rise construction might cast shadows on neighboring land. And accessibility for one is pollution or loss of security for others. Apart from potential ecological benefits, densification produces both advantages (e.g. increased housing options, and business opportunities) and disadvantages (e.g. rising noise or rents through costly upgrading of settlements, loss of green surfaces, or view) for different individuals, firms, or households.

This book identifies the socio-political challenges of implementing densification objectives, rather than considering the process as a technological, architectural, or design-based problem (see previous densification literature e.g. Bibby et al., 2018; Broitman & Koomen, 2015; Kyttä et al., 2013). The point of departure is that densification *per se* does not necessarily lead to sustainable outcomes in terms of social inclusion or community stability. Rather, how it is planned, implemented, and governed by the actors involved is what matters. Stated another way, the research gap this book addresses is to **politicize densification**. This is done by identifying the actors involved, their objectives, their strategies, as well as the socio-political structures (i.e. rules, laws, policies) that govern densification that try to prevent rivalries among competing groups and unsustainable social outcomes in housing, such as exclusion, gentrification, discrimination, or displacement.

## 1.5 Research Objectives and Questions

The overarching goal of this book is to examine, both theoretically and empirically, the different mechanisms that govern the implementation of densification objectives and its impact on housing uses, actor's strategies, as well as the impact densification has on social sustainability in housing (see Sect. 3 for more details).

*Theoretically*, this book aims to contribute to *actors-centered neoinstitutionalist political ecology* research (see Section 2) by analyzing how different (public and private) actors govern densification with regard to housing, and by focusing clearly on the social dimension of housing development. In addition, this book more than ever before aims to connect housing challenges to densification and land policy debates (e.g. Davy, 2012; Kolocek, 2017), as well as to social sustainability concerns (e.g. Bramley et al., 2009; Burton, 2000; Chiu, 2004). Considering future challenges of land scarcity that currently evolve in many cities globally, the findings of this research may help governments, practitioners, and planning professionals to cope with rising rental prices, exclusion, displacement, and social challenges in cities. Understanding the conditions for the success or failure of socially (un) sustainable implementations of densification objectives is an important step to overcoming barriers and to supporting policymakers and planning practitioners who promote more socially inclusive outcomes.

*Empirically*, research questions (see below) are answered by adopting a qualitative research design that is able to capture the origins of socio-political structures, human behavior, and decision-making (Sect. 4). Empirical research is done in Switzerland, a country that has been strongly challenged by rising housing use conflicts as results of densification in recent years, particularly in cities. While for the past twenty years much quantitative research in Switzerland has been employed to measure the impacts of urban sprawl (e.g. Grams & Nebel, 2013; Schwick et al., 2012), specifically on the quantitative effects of certain policy measures such as urban growth boundaries (e.g. Gennaio et al., 2009; Klaus, 2019; Weilenmann et al., 2017), this book aims to contribute to the work of fellow scholars who endeavored to analyze densification as a governance-oriented challenge from a qualitative research perspective (Balmer & Gerber, 2017; Devecchi, 2016; Hengstermann, 2019; Hersperger et al., 2014; Nabielek, 2011; Nicol, 2013; Nicol & Knoepfel, 2008; Rérat, 2012; Rudolf et al., 2018).

One overarching and three analytical research questions underlie this book. They are explained in more detail in the chapters that follow (Figure 1.1).

**RQ: What governance mechanisms lead to socially sustainable housing development in a densifying city?**

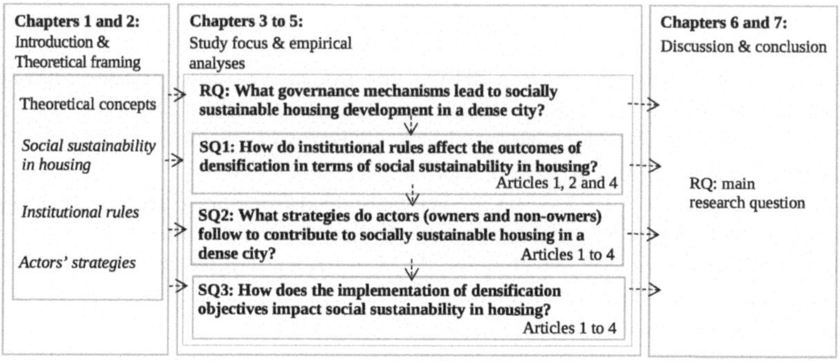


**Fig. 1.1** Schematic overview of the different elements and sections of this book

*SQ1: How do institutional rules affect the outcomes of densification in terms of social sustainability in housing?*
*SQ2: What use strategies do actors (owners and non-owners) follow to contribute to socially sustainable housing in a densifying city?*
*SQ3: How does the implementation of densification objectives impact socially sustainable housing outcomes?*

In the following sections, the theoretical background and analytical framework that I used in this book to answer the research questions are introduced (Chapter 2). In Chapter 3, the value added by the neoinstitutionalist political ecology approach is explained for the study of social challenges in housing and in a context of densification. In Chapter 3, in addition, the theoretical concepts of the analytical framework with regard to housing are explained in more detail. Particularly, the three main theoretical concepts this book is built on are introduced—housing as a resource, institutions, and actors' strategies—that help to answer the research questions. As demonstrated in the sections to come, these three blocks cannot be separated from each other as it is exactly the relationship that binds them—the governance mechanisms—which provides valuable insights into how actors involved in densification procedures govern housing socially sustainably (Chapter 3.1 to 3.6). In Chapter 4, I then introduce the study design and the geographical context of the empirical analysis (Chapter 4) and describe the structure of the four articles constituting this book (Chapter 5).

## References

Aalbers, M., & Christophers, B. (2014). The housing question under capitalist political economies. *Housing, Theory and Society, 31*, 422–428.
Aalbers, M. B. (2017). The variegated financialization of housing. *International Journal of Urban and Regional Research, 41*, 542–554.
Balmer, I., & Gerber, J.-D. (2017). Why are housing cooperatives successful? Insights from Swiss Affordable Housing Policy. *Housing Studies, 33*, 361–385.
Bibby, P., Henneberry, J., & Halleux, J.-M. (2018). Under the radar? "Soft" residential densification in England, 2001–2011. *Environment and Planning B: Urban Analytics and City Science, 47*(1), 102–118.
Boyko, C. T., & Cooper, R. (2011). Clarifying and re-conceptualising density. *Progress in Planning, 76*(1), 1–61.

Bramley, G., Dempsey, N., Power, S., Brown, C., & Watkins, D. (2009). Social sustainability and urban form: Evidence from five British cities. *Environment and Planning A, 41*(9), 2125–2142.

Breheny, M. (1997). Urban compaction: Feasible and acceptable? *Cities, 14*(4), 209–217.

Broitman, D., & Koomen, E. (2015). Residential density change: Densification and urban expansion. *Computers, Environment and Urban Systems, 54*, 32–46.

Burton, E. (2000). The compact city: Just or just compact? A preliminary analysis. *Urban Studies, 37*(11), 1969–2006.

Cernea, M. M. (1993). Culture and organization—The social sustainability of induced development. *Sustainable Development, 1*(2), 18–29.

Chiu, R. L. H. (2003). Social sustainability, sustainable development and housing development: The experience of Hong Kong. In R. Forrest & J. Lee (Eds.), *Housing and social change. East-West perspectives*. Routledge.

Chiu, R. L. H. (2004). Socio-cultural sustainability of housing: A conceptual exploration. *Housing, Theory and Society, 21*(2), 65–76.

Daneshpour, A., & Shakibamanesh, A. (2011). Compact city; Does it create an obligatory context for urban sustainability? *International Journal of Architectural Engineering & Urban Planning, 21*(2), 9–23.

Davidson, M., & Lees, L. (2005). New-build "gentrification" and London's riverside renaissance. *Environment and Planning A, 37*(7), 1165–1190.

Davy, B. (2012). *Land policy: Planning and the spatial consequences of property*. Routledge.

Devecchi, L. U. (2016). *Zwischenstadtland Schweiz—Zur politischen Steuerung der suburbanen Entwicklung in Schweizer Gemeinden*. transcript Verlag.

Dieleman, F., & Wegener, M. (2004). Compact city and urban sprawl. *Built Environment, 30*(4), 308–323.

Elkin, T., McLaren, D., & Hillman, M. (1991). *Reviving the city: Towards sustainable urban development*. Friends of the Earth.

Ewing, R. (1997). Is Los Angeles-style sprawl desirable? *Journal of the American Planning Association, 63*(1), 107–126.

Filion, P. (2015). Suburban inertia: The entrenchment of dispersed suburbanism. *International Journal of Urban and Regional Research, 39*(3), 633–640.

Freeman, L. (2001). The effects of sprawl on neighborhood social ties. *Journal of the American Planning Association, 67*(1), 69–77.

Frey, H. (1999). *Designing the city. Towards a more sustainable form*. E&FN Spon.

Gennaio, M. P., Hersperger, A. M., & Bürgi, M. (2009). Containing urban sprawl-evaluating effectiveness of urban growth boundaries set by the Swiss Land Use Plan. *Land Use Policy, 26*, 224–232.

Gordon, P., & Richardson, H. W. (1997). Are compact cities a desireable planning goal? *Journal of the American Planning Association, 63*(1), 95–106.

Gosling, V. K. (2008). Regenerating communities: Women's experiences of urban regeneration. *Urban Studies, 45*(3), 607–626.

Grams, A., & Nebel, R. (2013). Nutzungsreserven und -potenziale für die Siedlungsentwicklung nach innen. *disP—The Planning Review, 49*(2), 28–38.

Hackworth, J., & Smith, N. (2001). The changing state of gentrification. *Royal Dutch Geographical Society KNAG, 6*(9), 464–477.

Hengstermann, A. (2019). *Von passiver Bodennutzungsplanung zu aktiver Bodenpolitik—Eine Fallstudie zur Wirksamkeit von bodenpolitischen Instrumenten anhand von Lebensmittel-Discountern* (1st ed.). Springer VS.

Hersperger, A. M., Gennaio Franscini, M.-P., & Kübler, D. (2014). Actors, decisions and policy changes in local urbanization. *European Planning Studies, 22*(6), 1301–1319.

Holden, E. (2004). Ecological footprints and sustainable urban form. *Journal of Housing and the Built Environment, 19*(1), 91–109.

Holman, N., Mace, A., Paccoud, A., & Sundaresan, J. (2015). Coordinating density; working through conviction, suspicion and pragmatism. *Progress in Planning, 101*, 1–38.

Jenks, M., Burton, E., & Williams, K. (1996). *The compact city: A sustainable urban form?* (1st ed.). Routledge.

Kahn, M. E. (2000). The environmental impact of suburbanization. *Journal of Policy Analysis and Management, 19*(4), 569–586.

Klaus, J. (2019). Do municipal autonomy and institutional fragmentation stand in the way of antisprawl policies? A qualitative comparative analysis of Swiss cantons. *Environment and Planning B: Urban Analytics and City Science*, 1–17.

Kolocek, M. (2017). *The human right to housing in the face of land policy and social citizenship: A global discourse analysis*. Palgrave Macmillan.

Kyttä, M., Broberg, A., Tzoulas, T., & Snabb, K. (2013). Towards contextually sensitive urban densification: Location-based softGIS knowledge revealing perceived residential environmental quality. *Landscape and Urban Planning, 113*, 30–46.

Law, C. M. (2002). *Urban tourism: The visitor economy and the growth of large cities*. Continuum.

Lees, L. (2000). A reappraisal of gentrification: Towards a "geography of gentrification." *Progress in Human Geography, 24*(3), 389–408.

Lees, L. (2008). Gentrification and social mixing: Towards an inclusive urban renaissance? *Urban Studies, 45*(12), 2449–2470.

Marcuse, P. (1985). *To control gentrification: Anti-displacement zoning and planning for stable residential districts, Yearbook of construction articles* (pp. 931–952). New York University Review of Law and Social Change.

Nabielek, K. (2011). Urban densification in the Netherlands: National spatial policy and empirical research of recent developments. In *International forum on urbanism* (pp. 1–15).
Neuman, M. (2005). The compact city fallacy. *Journal of Planning Education and Research, 25*(11), 11–26.
Newman, P., & Kenworthy, J. (1999). *Sustainabily and cities: Overcoming automobile dependence*. Island Press.
Nicol, L. A. (2013). *Sustainable collective housing. Policy and practice for multi-family dwellings*. Routledge.
Nicol, L. A., & Knoepfel, P. (2008). Institutional regimes for sustainable collective housing stocks. *Swiss Political Science Review, 14*(1), 157–180.
Polese, M., & Stren, R. (2000). *The social sustainability of cities. Diversity and the management of change* (M. Polese & R. Stren, Eds.). University of Toronto Press.
Rérat, P. (2012). Housing, the compact city and sustainable development: Some insights from recent urban trends in Switzerland. *International Journal of Housing Policy, 12*(2), 115–136.
Rolnik, R. (2013). Late neoliberalism: The financialization of homeownership and housing rights. *International Journal of Urban and Regional Research, 37*(3), 1058–1066.
Rudolf, S. C., Kienast, F., & Hersperger, A. M. (2018). Planning for compact urban forms: Local growth-management approaches and their evolution over time. *Journal of Environmental Planning and Management, 61*(3), 474–492.
Schwick, C., Jaeger, J. A. G., & Bertiller, R. (2012). *Urban sprawl in Switzerland—Unstoppable? Quantitative analysis 1935 to 2002 and implications for regional planning*. Haupt Verlag.
Slater, T. (2009). Missing Marcuse: On gentrification and displacement. *City, 13*(2–3), 293–311.
Squires, G. D. (2002). *Urban sprawl: Causes, consequences and policy responses*. The Urban Institute Press.
Touati-Morel, A. (2015). Hard and soft densification policies in the Paris city-region. *International Journal of Urban and Regional Research, 39*(3), 603–612.
UN. (2018). *World urbanization prospects—The 2018 revision*. UN.
Weilenmann, B., Seidl, I., & Schulz, T. (2017). The socio-economic determinants of urban sprawl between 1980 and 2010 in Switzerland. *Landscape and Urban Planning, 157*, 468–482.
Whitehead, M. (2003). (Re)analysing the sustainable city: Nature, urbanisation and the regulation of socio-environmental relations in the UK. *Urban Studies, 40*(7), 1183–1206.

Wicki, M., Hofer, K., & Kaufmann, D. (2022). Planning instruments enhance the acceptance of urban densification. *PNAS, 119*(38), 1–10. https://doi.org/10.1073/pnas.2201780119

Williams, K. (2010). Sustainable cities: Research and practice challenges. *International Journal of Urban Sustainable Development, 1*, 128–133.

Williams, K., Burton, E., & Jenks, M. (2000). *Achieving sustainable urban form* (Williams, K., Burton, E., & M. Jenks, Eds.). E&FN Spon Press.

CHAPTER 2

# Governing by New Institutionalist Political Ecology

## 2.1 *Part I*: Theoretical Approach—Actors-Centered New Institutionalist Political Ecology

To understand the process of implementing densification goals in-depth and its socio-political consequences on different categories of stakeholders and their responses, the theoretical lens employed in this book applies an *actors-centered new institutionalist political ecology* approach. This approach builds a bridge between several disciplines with a spatial interest, in particular public policy analysis (planning as a public policy), new institutional economics (property rights), and political ecology (power).

By combining these perspectives, this approach makes it possible to recognize why many different stakeholders can come into conflict with each other. It allows for a more systematic analysis to examine how various actors behave in response to a specific socio-political setting and proves to be particularly suitable for the analysis of joint use situations in which several different users find themselves as rivals (such as in densifying urban environments). Moreover, it enables to systematically capture power games among actors to explain the function and evolution of structures that drive conflict and socio-environmental disputes in the first place. Hence, it allows us to reveal potential loopholes or challenges of densification implementation in more detail.

Regarding the compact city, more precisely, this new institutionalist political ecology approach acknowledges—for instance—that different



G. Debrunner, *The Business of Densification*,
https://doi.org/10.1007/978-3-031-49014-9_2

actors involved (e.g. owners, public authorities, NGOs, residents) try to shape the implementation of densification. Thus, densification objectives never get implemented on a one-to-one basis. Rather power games influence the way their implementation is or can be realized. Veto rights controlled by powerful stakeholders, as well as the negotiation of intertwined private and public interests, influence the way densification is being performed. Those actors simultaneously aim to defend their own interests and objectives, e.g. to preserve affordable living space, to save or consume energy, or to invest capital.

The shift towards the compact city increases the potential for use conflicts among these actors as it implies that they deal with the already built environment within firmly established city boundaries. Each one follows different strategies to defend their own interests and to achieve their goals of resource use in densifying city areas. In the chapters to come, densification is thus regarded and conceptualized as a highly sociopolitical challenge because its implementation results in losses for some and wins for others.

### *2.1.1 From Classic to New Institutionalism: Historical Background of Institutional Thought*

To understand what socio-political structures challenge the implementation of densification objectives, this book largely builds on theories and concepts of new institutionalism. Historically, institutional thought has a rich and diversified history in the social sciences (for discussion, e.g. Dembski & Salet, 2010; Immergut, 1998; Mandelbaum, 1985; March & Olsen, 1989). In general, an institutional analysis approach makes it possible to explain human behavior as results of joint values, norms, routines, and procedures stipulated in formal rules, codes, and ordinances that guide social behavior and action. Institutions are defined as "the conventions, norms, and formally sanctioned rules of a society. They provide expectations, stability and meaning essential to human existence and coordination. Institutions regularize life, support values and protect and produce interests" (Vatn, 2005: 24).[1] They range from the rules of a constitutional order to the standard operating procedures of a

[1] Similar definitions of "institutions" have also been introduced by Ostrom (2007: 22) und North (1991: 91).

bureaucracy or firm relations. Classic institutionalists regard the institutional organization as the principal factor structuring collective behavior in society and in generating distinctive outcomes (Hall & Taylor, 1996: 937). Following this approach facilitates an understanding of *how* densification as a process is embedded into diverse institutional structures (laws and policies) that influence actors' decision-making behavior.

By the end of the 1980s, scholars of several sub-disciplines of political science studies, economics, and sociology rediscovered, quite independently from each other, the potential of this approach and began talking of a "new institutionalism" (for discussion, e.g. Koelble, 1995). In contrast to classic institutionalism (see previous paragraph), which often led to unraveling the functioning of institutions in a descriptive and legalistic language (Thelen, 2003), new institutionalism or neoinstitutionalism as a theoretical concept developed from the behavioral, cultural, and spatial turn during the 1960s and 1970s. The main purpose of this turn was to elaborate upon the role of institutions in the determination of social and political outcomes in more detail (Hall & Taylor, 1996: 936).

In particular, neoinstitutionalists have started to acknowledge that it is exceptionally relevant to analyze the key attributes of human action for understanding sustainable development. Different categories of institutional rules have become considered necessary for understanding the outcomes of social behavior and practice. Simultaneously, institutions themselves have come to be understood as a product of social construction rather than given per se (e.g. Williamson, 2000). In contrast to classic institutionalism, new institutionalism goes one step further in the sense that it raises new questions: for instance, *why* institutions have emerged the way they did. It also focuses more closely on microanalytic perspectives and criticizes the image of social causation as "path dependent", while respecting that the effects of institutions are mediated by contextual features of a given socio-political situation and are often inherited from the past (Hall & Taylor, 1996; Healey, 1996, 1999; Powell & DiMaggio, 1991). New institutionalists have also started to emphasize in a more detailed manner that, besides the importance of public policies, property rights (e.g. Demsetz, 1967; Jacobs & Paulsen, 2009), rent-seeking, and transaction costs (costs other than those involved in the physical production of buildings) (Coase, 1960) play a crucial role in the operation and development of institutions. In Sect. 3.4, particular emphasis is put on the different forms of institutional rules influencing densification implementation.

### 2.1.2 From Hardin to Ostrom: Towards Political and Actor-Centered Neoinstitutionalism

In this book, densification procedures in the urban housing sector are analyzed by applying theoretical concepts deriving from political and actor-centered neoinstitutionalism (e.g. Mayntz & Scharpf, 1995; Scharpf, 1997, 2000). This approach has received its popularity with the Nobel-Memorial-Prize-winning[2] political scientist Elinor Ostrom in 2009. Ostrom won the award in economics because she disapproved of Garret Hardin's concept of "The Tragedy of the Commons" (1968). In his metaphor, Hardin argued that a common good such as land for cattle could not be used in a sustainable way, as finally, this would lead to overuse and unproductivity of the land. He saw this observation as an irrefutable argument for the superior efficiency of private property rights in the management of land and other common pool resources (CPRs) such as air or water as well as an undeniable justification for privatization.[3]

Elinor Ostrom, however, countered some of Hardin's presumptions in her book *Governing the Commons* (1990) and showed that it is, in fact, private property in cattle and individual utility maximizing behavior that lies at the heart of the problem rather than the common property character of the resource. To prove her assumption, she analyzed the sociological, historical, and anthropological structures that guide the use of natural resources and showed that, if the herders talked to each other, or shared cultural customs and procedures, they might be able to solve any commons challenge. She showed that individuals are capable of developing sensible collective ways to manage common property resources for

[2] The Memorial prize is donated by the national bank of Sweden [Sveriges Riksbank] in economic sciences in memory of Alfred Nobel. This has caused much controversy among scientists as the price is not awarded by an academic organization rather than connected to the bank's own ideology.

[3] In 1991, Hardin published another study on the subject because he felt himself misunderstood. He corrected his statement from "The Tragedy of the Commons "by not exclusively defending private property rights, but rather unregulated private property rights (Hardin, 1991). In particular, he argues that the tragedy applies to open-access commons, and that privatization or top-down regulation (restricting access) is necessary to avoid depletion. In contrast, Ostrom showed that there are many examples of self-governance in which communities sustainably used commons without top-down regulation or privatization. In her perspective, it is more about how and whether bottom-up and self-governance can lead to sustainable management of commons.

individual and collective benefit. Her main concern was then to investigate *how* and *why* in some cases stakeholders succeed in doing so and under what conditions they do not.

With the findings of her research, she questioned the long-lasting economic orthodoxy, which only recognized policy in terms of a dichotomous choice between state and market, but not as a stand-alone and integral part of the socio-economic system that determines the use of sustainable resources (Harvey, 2012: 191–207). Her results have led to the recognition and integration of more cultural approaches of sociology, anthropology, and philosophy in political economy (e.g. Berger & Luckmann, 1966; Bourdieu, 1977, 1998; Weber, 1968). Scholars of this research field have started to (re)consider social norms and their guidance function over action, in addition to purely formal rules and procedures.

Indeed, Elinor Ostrom was one of the first to describe, analyze, and explain environmental problems and the unsustainability of natural resources use as results of institutional patterns *and* the involved actors' individual behavior. She fundamentally questioned the dichotomy between the state and the private sphere and discussed the potential of other forms of use rights to regulate resources (e.g. public or collective). William Blomquist (2012: 370) concluded that Ostrom was a pioneer in raising unique questions such as "how people create property rights and for what purposes, why and how they choose the types of property rights institutions they do, and how and why they change property rights over time. [...] She was explicitly and primarily concerned with (1) 'who gets what, when, how' to quote Lasswell's (1936) famous characterization of politics; and (2) the even more intensely political questions of who decides who gets what, when, and how, and how that question is decided". Ostrom accepted the fact that political use conflicts among rival groups lie at the heart of resource scarcity and raised awareness that environmental concerns are ultimately political problems (Bookchin, 1993).

Her work inspired many different scholars from other fields (e.g. anthropology, ecology, sociology), and particularly led to new debates and criticism in neoclassical economy. Connected to the writing of Elinor Ostrom, the award-winning economist Douglass C. North (1990, 1994), for instance, argued that neoclassical theory (e.g. Becker, 1976; Schumacher, 1973) failed as it does *not* consider political structures as a fundamental element explaining economic and environmental change. Neoclassical economics held the environment separate from humans and their economic activities. Therefore, the origins of resources depletion cannot be explained in-depth because markets are not recognized as results of socio-political structures and human action.

By the end of the 1970s, the research field—of which Ostrom and North became the most prominent representatives—was summarized under the new approach and term "New Institutional Economics". This new approach "entails trade-offs between environment and development and the integration of the economy and the environment. Under the balance rubric, environmental justice, economic equity, and other manifestations of redistributive justice draw their basis" (Neuman, 2005: 19). The central message of new institutional economics is that institutions matter for economic performance. Because resource scarcity is mediated through institutions, it is acknowledged to be politically and socially constructed (Shahab et al., 2019: 541). As a refinement of the neoclassical model, new institutional economics regard the way by which property rights are allocated and enforced as determined by transaction costs because any kind of economic exchange results in external effects that need to be internalized by incentives. Changes from common to exclusive private property rights therefore leave room for unexploited gains of exchanges or benefits. While neoclassical economics consider this change in property agreements as triggered by self-interest and as results of spontaneous order, new institutional economists emphasize that they are imposed on society by civil authority, the state, and in the interests of individuals.

Later on, Elinor Ostrom incorporated her findings into the "Institutional Analysis and Development" (IAD) framework, which was used to systematically analyze policy processes and outcomes in the study of common goods such as fishery stocks or woodlands (Ostrom, 2005, 2007). According to Ostrom, the IAD helps to understand how governance systems enable individuals to solve specific problems democratically and how to organize diagnostic and prescriptive capabilities. Without the use of a framework, the systematic and comparative institutional assessment would not be based on analysis of performance, but instead on normative ideas about what kinds of institutions are "good" or "bad" (Ostrom, 2011: 7). Even though in this book the IAD is not applied, I agree that an analysis framework is needed to analyze densification processes systematically. In the following chapter, therefore, the Institutional Resource Regime (IRR) analytical framework is introduced, which allows me to address the research questions in an appropriate manner.

### 2.1.3 The Question of Power: Bringing a Political Ecology Perspective to Neoinstitutionalist Research

*Political ecology* is a field of research within socio-environmental studies. Its core endeavor is its focus on power relations, socio-economic and political processes as well as the coproduction of nature and society (Swyngedouw, 2009). It moreover focuses on the study of the relationships between political, economic, and social factors that cause environmental degradation (Robbins, 2004), *and* the reasons why land use conflicts arise in the first place (e.g. Lasswell, 1936; Ribot & Peluso, 2003; Robbins, 2004; Swyngedouw, 2009). Theoretical inspirations to explain environmental changes are taken from different sources grounded in human geography, social anthropology, development studies, or heterodox economics.

In contrast to apolitical explanations of environmental change (e.g. Neo-Malthusianism, Limits to Growth models), political ecology differs by politicizing environmental issues and phenomena. It demonstrates that natural resources use is linked with distributive political processes, practices, and power asymmetries among different groups of stakeholders (Gerber & Debrunner, 2021). Identification occurs by highlighting questions of power, responsibility, decision-making, capacity-building, and sustainability (e.g. Evans & Jones, 2008; Evans et al., 2006; Haller et al., 2016; Krueger & Agyeman, 2005). These authors provide an alternative view on mainstream environmental degradation discourse that often puts the blame on local communities or on the least well off from society in general. Moreover, these authors question the activity of resource exploitation that powerful economic actors engage in, and they make "explicit considerations of relations of power" (Robbins, 2004: 12).

In addition, the approach takes consideration of the political processes through which resource access is defined, negotiated, and contested at multiple scales (Zimmerer & Bassett, 2003). The status of powerful actors (e.g. conservation organizations, governments, businesses) and what is taken for granted in leading discourses is critically and explicitly questioned and reflected upon (Svarstad et al., 2018). Although political ecology approaches have so far mainly focused on environmental or land use change in the Global South, challenges in first world political ecology (such as the policy shift towards densification) demonstrate that political ecology perspectives also concern the Global North (McCarthy, 2002).

A useful definition of the school of thought can be found in Robbins (2004: 12), who describes political ecology as the attempt to search for "empirical, research-based explorations to explain linkages in the condition and change of social/environmental systems, with explicit consideration of relations of power. Political ecology, moreover, explores these social and environmental changes with a normative understanding that there are very likely better, less coercive, less exploitative and more sustainable ways of doing things". Despite apparent agreement that *power* is at the core of political ecology studies, over the years diverse power definitions have been introduced. A popular definition was given by Ribot and Peluso (2003: 156), who defined power, first, as "the capacity of some actors to affect the practices and ideas of others [...], and second, [that] power [is] emergent from, though not always attached to, people. [...] Disciplining institutions and practices can cause people to act in certain ways without any apparent coercion". Power is studied as contestations over material assets (land, natural resources) as well as over meaning (Benjaminsen & Svarstad, 2019: 391).

In this book, I focus on a particular political ecology perspective that emerged from the 1970s as a result of a Neo-Marxist critique of Malthusian ideas in environmental thinking (also the IRR analytical framework developed within this tradition) (see e.g. Ehrlich, 1986; Enzensberger, 1974). The argument was that studies of human ecology are never neutral or apolitical, but involve interests, norms, *and* power. In line with this understanding, *power* is then seen as "the ability to get what one wants from others. It may come from greater wealth or social position or the ability to manipulate the ideology of others" (Ensminger, 1992: 7). Not only are powerful individuals more likely to influence institutions to their own advantage, but also "any given set of rules or expectations - formal or informal - that patterns action will have unequal implications for resource allocation, and clearly many formal institutions are specifically intended to distribute resources to particular kinds of actors and not to others" (Mahoney & Thelen, 2010: 8). Powerful actors are those who know how to influence the goals of the others in a targeted manner to promote or to protect their own values, needs, and objectives. Vice versa, the institutional framework also shapes and affects all actors' behaviors, as well as their negotiating power (Mackay et al., 2010).

Environmental degradation was and still is seen as caused by human impact and political choices creating winners and losers. In particular, Neo-Marxist political ecologists address the issue of power by relying on a political economy perspective (for detailed discussion see Haller, 2019; Svarstad et al., 2018) that insists on the need to link the distribution of power with productive activity and ecological analysis (Robbins, 2004). Such perspective points out the role of market (capitalistic) relationships in the stratification of society with differential distribution of power among social classes, and corresponding impact on resource use and overuses (Robbins, 2004). The study of power relations is connected to the question of *who* controls the access to and uses of natural resources. Whether through exclusive property rights or tenure arrangements or through mechanisms of social exclusion from decision-making (Ribot & Peluso, 2003). Hence, Neo-Marxist political ecology underlines the agency of resource users and aims to analyze the conditions triggering forms of resistance against more powerful actors following productivist objectives in the management of resources (Haller, 2019). It points out that ecological objectives should not be the starting point of analyses of resource degradation and sustainability, but socio-political conflicts—which are often unspoken environmental conflicts (Martínez-Alier, 2002)—targeting institutional change.

This book adds the above-mentioned political ecology perspective to neoinstitutionalist research (this chapter) as it not only endeavors to describe and to analyze socio-political phenomena, but also aims to explain and to critically question them. Vice versa, political ecology does not manage to identify power structures in any systematic way. Even though it might help to ask the right questions, it does not offer any assistance in providing concrete answers to them (Hengstermann, 2019: 8). Therefore, in this book, I aim to open this dialogue between new institutionalism and political ecology to insufflate greater power-awareness to neoinstitutionalist research. Power relations that drive densification are identified and systematically explained. In this chapter, Part II, I will continue to present the core elements of the Institutional Resource Regime (IRR) analytical approach in greater detail—especially how it is applied to the study objective of housing development in dense urban environments (Chapter 3).

## 2.2 *Part II*: Analytical Framework—The Institutional Resource Regime (IRR) and Its Focus on Property Rights

Based on the major conceptual shift that took place with the evolved approaches of new institutional economics (North, 1990, 1994; Ostrom, 1990), the Institutional Resource Regime (IRR) analytical framework has emerged. This particular neoinstitutional analysis approach combines new institutional economics *and* property rights theory with policy analysis (for discussion, Aubin, 2007; Gerber et al., 2009, 2020; Knoepfel et al., 2001, 2003). The IRR framework is rooted in political science or, to be more precise, environmental policy analysis. It enables "a systematic analysis of the institutional context that influences actor behaviour and the use of natural resources" (de Buren, 2015: 9). It moreover postulates that a combination of approaches from political science (in particular policy analysis) and institutional economics (of property rights) ensures the identification of the most relevant institutional dimensions, which can explain the (un)sustainable use of resources (Gerber et al., 2009: 799). Thereby, distinct insights into the diverse array of regulatory conditions and actors are provided.

Other than Ostrom's IAD framework, the IRR explicitly distinguishes between two main sources of formal rules—public policies (Sect. 2.2.2) *and* property rights (Sect. 2.2.3)—which simultaneously influence the use and disposal rights of resource use and have very different characteristics (Gerber & Nahrath, 2013: 12). The IRR does not only focus on single use situations with model character (e.g. pastures for cattle) or analyze a limited number of actors, but it rather follows a more comprehensive approach. It proves to be particularly suitable for the analysis of joint use situations in which several different users find themselves as rivals (such as in dense urban environments) because it is able to fully take into account the role of the public state (Knoepfel et al., 2003).

In contrast to the IAD, the IRR framework facilitates an understanding of how various actors behave in response not only to changes in the individual regulations of the institutional setting but also to various characteristics of the institutional rules involved as a whole (Nicol, 2011: 460). To identify conflicts that arise during contested use situations, which potentially hamper a resource's sustainability, the IRR examines in-depth *how* and *why* rival use situations between multiple actors emerge. This high complexity of user situations helps "to understand a more representative range of resources uses [in order to] be capable of portraying the complexity of heterogeneous use situations" (Gerber et al., 2009: 800). In this book, the IRR's ability to analyze complex use situations becomes particularly relevant since densification potentially exacerbates resources use conflicts among actors due to limited land availability.

In sum, this approach is to be seen as the next generation of institutional analysis as it incorporates earlier models, such as Ostrom's IAD (2007, 2009). It allows for analyzing behavioral patterns stemming from incentives of different policy fields such as contradictions between public policies (e.g. planning law) *and* property rights (e.g. Civil Code). Furthermore, a major strength of the IRR framework is its ability to conceptualize institutions in a way that echoes real-life resource use situations by taking their complexity into consideration and to propose causal mechanisms explaining the relationship between institutions *and* sustainability. The IRR therefore contributes to a broad set of questions on the political and institutional dimensions of resource governance (Gerber et al., 2020).

It must also be noted that Ostrom's IAD originated in the context of the Anglo-Saxon legal system (common law countries) while the IRR has evolved in countries with a codified legal system (civil law countries). This further presents an added value of the analytical framework applied in this book.

### 2.2.1 *The Institutional Regime: Two Sources of Formal Rules Regulating Resources Use*

As mentioned before, the IRR distinguishes two main categories of formalized rules—public policies *and* property rights—that operate according to a different logic and rely on opposing legitimizations. These two sources of institutional rules form an Institutional Regime that regulates resources uses. A regime is understood as "the more or less coordinated combination of public policies and property rights that relate to all user-actors of the resource, and thus affects the reproductive capacity of the resource and hence its sustainability" (Nicol & Knoepfel, 2008: 161). An institutional regime defines an explicit (or implicit) structure of rights and duties, characterizing the relationship of individuals to one another with respect to a particular resource such as land or housing (Bromley, 1992: 8).

A regime creates exclusivity, adumbrates and demarcates how resources can be used by whom and sets the rules as to how various actors can gain access, use, or exploit them. The institutional regime also manipulates, restricts, or enhances actors' use interests (Gerber et al., 2018: 3). Regime types can be defined and categorized on the basis of two dimensions—*extent* (quantitative dimension) and *coherence* (qualitative dimension) (Hengstermann, 2019: 101). Extent refers to the total number of regulations that influence the different uses of a resource at a given time (e.g. energy law, housing law, etc.). The criterion of coherence depends on the level of content and connection *between* public policies *and* property rights (external coherence), but also *within* public policies *or* property rights (internal coherence) (Gerber et al., 2009: 8). The IRR analytical approach assumes that the extent and the coherence of resources use are intrinsically linked "because any increase in the number of regulations tends to generate inconsistencies. Conversely, when only a few uses are regulated, the coherence is likely to be much greater" (de Buren, 2015: 16).

Use situations in which the extent of regulations leads to internal or external incoherence are called "complex regimes". In this situation, the majority of the goods and services provided by the resource are actually regulated, but in a way that is incoherent in part. The risk of over-exploitation of resources increases. This situation corresponds to many resource regimes in liberal states because of the extensive development of largely uncoordinated policies since the 1950s (Gerber et al., 2009: 8). In contrast, a resource regime in which extent and coherence are balanced is likelier to be an "integrated" regime. Then, all goods and services produced by a resource are regulated in a coherent way, which increases the possibility that sustainable use conditions are created. The central and overarching assumption of the IRR analytical framework is that "high levels of regime extent and coherence are necessary preconditions for resources sustainability" (Gerber et al., 2009: 798). In states based on the rule of law, it is therefore necessary to perform a close analysis of the legal foundations of the legal system as well as of its functioning and characteristics to understand the institutional regime in force.

At this point, it must be noted that besides the analysis of formal institutions, the IRR also considers the impacts of informal rules such as social norms, conventions of social behavior, sanctions, or taboos. Informal institutions describe actors' norms in a given context and the ensuing generally accepted rules-in-use (Thomann et al., 2018). Informal rules appear and thrive in the interstices left between formal rules (e.g. de Buren, 2015; Gerber et al., 2020). The IRR analytical framework assumes that informal rules, sooner or later, result in formal institutions to a greater or lesser extent (Gerber et al., 2009: 803). While in situations of informality actors do *not* play by the formal rules they *do it against* the rules and thus, indirectly acknowledge them (Gerber et al., 2009). Consequently, informal rules have a lasting impact on the ways a society conducts itself. This means that, even in "weak states", the formal legal framework provides a strong reference that shapes individuals' actions (Hagmann & Hoehne, 2009). In any case, formalized institutions strongly and directly influence the use interests of all relevant stakeholder groups and embody the clearest expression of social will in dealing with a resource.

In the next section, I explain in more detail these two categories of formal regulation—public policies *and* property rights. *First*, I address how they constitute the institutional regime of a resource; *second*, how they influence the behavior of actors involved such as owners and other user actors; and *third*, how incoherencies in regulations might occur that produce unsustainable outcomes.

### *2.2.2 Public Policies*

According to Knoepfel (1986), the dialectical relationship between public policies and property rights in resource regulation can be summarized as follows: *Public policies* are crafted by democratically elected bodies to solve a politically defined public problem in the interest of the voting majority. Public policies are regularly revised and updated by political actors (Knoepfel & Nahrath, 2007: 24). Moreover, public policy derives from the state's attempt to solve what it considers a public problem and is expressed in the body of laws, regulations, decisions, and actions of the government (Nicol & Knoepfel, 2008: 170). Following the IRR approach, public policy hence entails "a series of intentionally coherent decisions or activities taken or carried out by different public and sometimes private actors whose resources, institutional links and interests vary, with a view to resolving in a targeted manner a problem defined politically as collective in nature" (Knoepfel & Nahrath, 2007: 24).

This problem gives rise to the introduction of formalized acts of a more or less restrictive nature that are often aimed at modifying the behavior of target groups (social groups presumed to be at the root of or able to solve the collective problem) in the interest of the social groups who suffer the negative effects of the problem in question (final beneficiaries) (Knoepfel & Nahrath, 2007). Hence, through public policies the state receives the power to regulate the actions of those actors who are thought to be at the source of the collective problem, in the name of public interest. Public policies are regularly revised not only because the collective problem they are targeting constantly evolves, but also because changing political majorities propose alternative solutions to the problem (Knoepfel et al., 2012: 417). In daily practice, this makes it sometimes difficult to enforce public interests in a targeted manner since public policies are not as stable and resistant as private property rights (Sect. 2.2.4).

Public policies, moreover, are enshrined in public laws such as housing laws, planning laws, or building laws and regulations (Fig. 2.1). These laws (e.g. Housing Acts, Planning Laws) provide the legal framework for implementing the policies and guiding the actions of governments, individuals, or organizations involved in housing and urban development. In addition, all of them simultaneously solve a public problem (e.g. urban sprawl) and draw its legitimacy from a legal basis (e.g. constitutional article). This legal basis not only defines the public actors in charge of implementation but also provides them with a budget, personnel resources, and so on (Gerber et al., 2017: 1687).

#### *2.2.2.1 Land Policy Instruments Steering Spatial Development*

Among such public policies is "land use planning policy" (or land policy), which is about creating or alleviating scarcity of land according to politically predefined spatial development objectives (Gerber et al., 2018). Land use policies govern how land (and other resources e.g. energy) is allocated, used, and distributed in order to steer spatial development and resources use in the interest of the public (Gerber et al., 2017). As a public policy, moreover, it draws its legitimacy from a legal basis (e.g. constitutional article, land use planning law), defines the public actors in charge of implementation (e.g. planning agency), and provides them with a budget, competencies, or means of action (Viallon, 2018). Land use planning has typically developed a series of hierarchical plans (policy instruments) to control the actions of those actors, namely the landowners, whose effective behavior has been identified as leading to uncoordinated growth. More specifically, land use policy instruments (or "tools " in the U.S. context) refer to specific institutional rules, laws, and

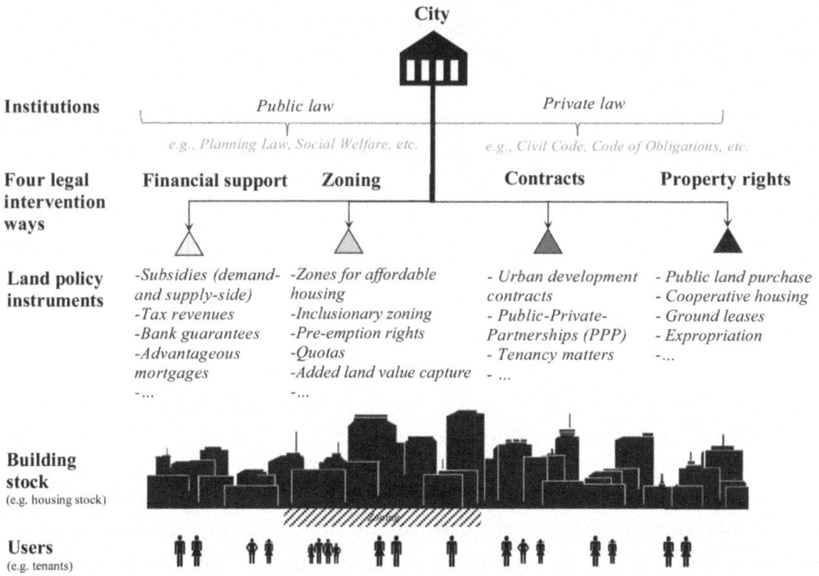


**Fig. 2.1** Instruments of land policy (own Figure based on Balmer & Gerber, 2017; Gerber et al., 2018: 16)

regulations—enshrined in public law *and* private law (see Chapter 3.4 of this book for more details)—aiming to shape and influence land use, development, management, distribution, or allocation, etc. In their book, *Instruments of Land Policy (2018)*, Jean-David Gerber, Thomas Hartmann, and Andreas Hengstermann provide a comprehensive overview of potential land rules in force (Fig. 2.1).

### 2.2.3 Property Rights—A Chronological Review

The IRR analytical approach acknowledges that besides public policies, property rights play a decisive role in the regulation of resources (for discussion, e.g. Demsetz, 1967; Jacobs & Paulsen, 2009; Steiger, 2006). In general, "property" does not describe an object such as land. Rather it is understood as a social relation that defines the property holder with respect to something of value (the benefit stream) against all others

(Hallowell, 1943: 115–138). The one holding said rights has the expectation, in both the law and in practice that, those with duty will respect his or her claims.

As titleholders, property owners have the legitimate authority to act in a predetermined manner. Thus, property rights are identified as an exclusive, transferable, and legal right to the physical use of scarce resources, the returns thereon, and the alienation thereof. The authority system legitimizing this behavior can either be a central government, or it can be a local village council. The important issue is that the individuals feel compelled to comply with the institution in effect (Bromley, 1991).

In liberal states, therefore, *property rights* protect the individual's interest against the (potentially absolutist) action of the state. Property rights are grounded in the Civil Code (or similar in Common Law countries) and are extremely stable over time because their definition hardly changes (Bromley, 1992: 11; Savini et al., 2015). Within the constraints of the law, the holder of a property right has the right to benefit as well as to freely and completely dispose of his or her property. For example, land or housing stock owners have (1) the right to control and make decisions about the housing stock that belongs to them; and (2) the right to obtain at least a portion of the benefits produced by the housing stock. Owners are also legally bound to fulfill certain obligations. The Code of Obligations and supplementary contracts stipulated in private law describe the obligations of the stock owners regarding, for instance, contractual obligations, the sale of buildings, rental contracts, or the relationship to tenants (Nicol & Knoepfel, 2008: 170).

Moreover, Bromley (1992: 2) distinguishes between four different property regimes that structure resources use—common, state, private, and non-property (open access). He argues that property regimes represent human artifacts reflecting instrumental origins that provide owners the legally and socially sanctioned ability to exclude certain users and to force them to go elsewhere (Bromley, 1992: 15). Consequently, particular property regimes prove to be chosen for particular purposes. For instance, concerning housing stock regulation, the importance of private property lies in the fact that the surrounding laws stipulate specific forms of tenure that enable residents' entry or exit to/from the housing estate. Private rental, public housing, or tenant cooperatives are very different forms of ownership that include the potential for social exclusion to a greater or lesser extent (Blomquist, 2012; Wimark et al., 2019: 20).

In the following section—still subsection to 2.6—two instrumental conceptions of private property are introduced. Historically, they have followed either classic liberal or reformist tradition. Both of them acknowledge private property as the most important source of economic activity (for discussion, e.g. Meyer, 2009). In this book, this chronological review helps to understand *how* property rights have evolved over time and *how* this background explains *why* property owners are able to defend their interests in dense urban environments in an effective manner in contrast to other user actors, such as public authorities or tenants.

#### *2.2.3.1 Property Rights in the Rise of Liberal Philosophy—Protection Against State Powers*

During the seventeenth century in the rise of liberal philosophy, the conception of landownership was based on the idea of "private dominium" (Hobbes, 1651; Locke, 1689). It was handed down from Roman law, revived by the Napoleonic Code after the French Revolution, and subsequently spread within Europe and throughout the world, particularly through colonization. Following the principle of accession, individuals started to exercise dominion over several things they owned. They also became the owners of immovable objects such as buildings that were attached to the land (for discussion, Gerber et al., 2017: 1687). By further tightening "the bundle of rights" (Commons, 1893: 263), the accession principle played a fundamental role in making property more exclusive and rigid. The metaphor of "the bundle of the rights" conveys that property rights are to some extent decomposable into elements that secure property owners' exclusive rights of resources use such as the Rights of Access, Rights of Withdrawal, Rights of Management, Rights of Exclusion, and the Rights of Alienation (Schlager & Ostrom, 1992: 252). These rights derive their significance from the fact that they help an owner to form those expectations that he or she can reasonably hold in dealings with others.

Specifically, in classic liberal perception, the purpose of property rights is to maximize the common good when socially integrated through the institutions of fair and free market exchange. The solution to the problem of natural monopoly is to conduct an *ex ante* bidding competition and award the right to serve the market to the group that tenders the best bid. Thomas Hobbes (1588–1679)—English liberal theorist and philosopher—argued that a commonwealth is only produced through privatizing competitive interests within a framework of strong state power. Private property was identified as an individual's right that arises when they create value by mixing their labor with the land. The fruits of their labor belong to them and to them alone. Market exchange socializes that right when each individual gets back the value they have created by exchanging it against an equivalent value created by another (Hobbes, 1651). In effect, individuals maintain, extend, and socialize their private property right through value-creation and supposedly free and fair market exchange.

Like Thomas Hobbes, John Locke (1632–1704) later on believed in a natural right to life, liberty, and property. However, according to his perspective, the right to property is guaranteed to protect individuals against possible interferences—from other individuals or the government—in the private sphere. The function of property is to protect individuals' autonomy and freedom as citizens (Locke, 1689). Property rights make the private appropriation of goods and services provided by resources possible, as long as public policies do not restrict exclusive appropriation in the name of general interests(s) (Constant, 1988). This classical economists' notion of property rights was articulated by many other liberal theorists such as Adam Smith (1776), Thomas Malthus (1789), or David Ricardo (1812), and found its ways into the state systems of most Western modern societies (Keynes, 1936).

In sum, this classical liberal conception of property has always been very strongly linked to capitalism. The credo was and still is that human well-being can best be advanced by liberating individual and entrepreneurial freedom and skills within an institutional framework characterized by strong private property rights, unencumbered markets, and free trade (for discussion, Miller, 1978). Political sovereignty appears as a necessary evil that intervenes in and violates the pre-existing private sphere of ownership (Meyer, 2009: 104). Private property can only be infringed upon in very particular circumstances and when "fair compensation" is guaranteed (Cooter & Ulen, 2004; Hartmann & Needham, 2012). The role of the state, in turn, is to create and to preserve an institutional setting appropriate to such practices (Harvey, 2005: 2–3).

#### *2.2.3.2 Reformist Positions—Towards a Perspective of Possession*

During the 1870s, the classic liberal perception of property was questioned by those who pointed out that titleholders also have a social responsibility (Commons, 1893; Engels, 1872; Marx, 1868; Proudhon, 1840). By the early twentieth century, urban growth, affordable housing crises, and severe hygiene problems have created reform movements.

These proposed that one of the missions of private property is to promote societal goals that are intrinsically associated with a social obligation for the landowners (Jacobs, 1998; Sax, 1992). In response to the acute housing crises endured by the working class in many cities at that time, Karl Marx and Friedrich Engels—two German followers of the French anarchist Pierre-Joseph Proudhon—proposed outlawing private landlordism and converting tenants' rents into purchase payments on their dwellings (Hodkinson, 2012). This, they believed, would end the exploitative character of private property and transform the property-less tenants into a "totality of independent and free owners of dwellings" (Engels, 1872/1997: 28).

Their proposal was based on the writings of Karl Marx (1859)—German philosopher, economist, and political theorist—who observed that the use of goods and services under liberalism was shaped by a process of "commodification". This process perpetually aims to produce surplus products in order to constantly gain a surplus in economic profit. Market forces and profitability objectives determine not only *how* goods and services are used, but also *how* they are produced, managed, and distributed (for discussion, e.g. Harloe, 1982: 40; Harvey, 2005: 166). While commodification might be advantageous for those selected few who reap the disproportionate benefits of the capital gain, the vast majority and particularly those of lower income would have little ability to capture value from this development. Commodification, therefore, would sooner or later end in economic development that is (ab)used as a source for profit extraction by a small financial elite. Those with limited financial means, however, will be pushed out and excluded from this process.

A sector in which "commodification" has become especially apparent was and still is the housing segment. According to Marx (1859), the commodification of housing relies on the assumption that the market, including the profit-maximizing rationality of investors, is the most efficient solution to guarantee housing supply for all income segments. The role played by private landowners becomes particularly relevant in this matter. Due to the protection guaranteed by private property rights, landowners are free to define the profit margin to be targeted on their parcels and to set the rents according to market prices. Hence, commodification of housing not only results in a dominance of financial actors, markets, practices, and narratives at various scales but in a structural transformation of housing supply, demand, and households itself (Aalbers, 2019: 4). It moreover leads to a general shift from housing treated by its use value to its financial value, meaning that housing is no longer considered a basic human need but rather more a commodity that must be traded or paid for in a globalized financial market (for discussion, e.g. Aalbers, 2017; Harvey, 1985; Marcuse, 2012; Rolnik, 2013; Schipper, 2014). Housing stocks, in other words, have become a lucrative investment outlet, a safe source of revenue, and a highly valued form of collateral.

Engels moreover argued that there was no such thing as a housing crisis, only a crisis of capitalism in which housing conditions formed just "one of the innumerable smaller, secondary evil" caused by the exploitation of workers by capital (Engels, 1872/1997: 18). The contradictory and uneven processes of capitalist development would, sooner or later, continue to generate housing questions at different points of the business cycle. From this observation derived one inescapable political conclusion: the only real alternative to the housing question was "to abolish altogether the exploitation and oppression of the working class by the ruling class (Engels, 1872/1997: 17) through working class revolution and expropriation of private property.

Later on, this classical Marxist orthodoxy that only a proletarian revolution would be able to solve housing challenges has sparred with less rigorous interventions. Other reformist positions aimed at reconsidering the strategic importance of state intervention or the shift towards more self-organized solutions of property in the here and now, such as small-scale cooperatives or mutual ownership (for discussion, Hodkinson, 2012). Without denying the commodity character of housing that Marx and Engels have pointed out, these reform approaches have brought to the surface the use value of housing both as an essential human activity and as a sphere of productive non-market activity.

### 2.2.4 *Conflict Relationship Between Public Policy and Property Rights*

Consequently, public policies with a spatial impact often conflict with the property owners' freedom since their rights might be restricted (e.g. through zoning). Among these public policies, land use planning is the most obvious as it precisely aims to control how landowners use their land and the housing stock that is built in the interest of the public. However, since private property rights are strongly protected by law and are very inflexible, land use planning seems to experience difficulty in implementing democratically accepted spatial development plans on titleholders due to conflicting interests (Jacobs & Paulsen, 2009; Needham & Verhage, 1998). For example, densification efforts to curb urban sprawl prove very tricky to implement (see next paragraph). Therefore, the real spatial challenge is not so much plan making, but rather plan implementation. Without heavy state intervention such as expropriation, new housing regulations (e.g. new zoning) are only implemented when titleholders agree to undertake new developments, sell their stock or the land, or transfer their development rights (Gerber et al., 2017: 1685).

The shift towards densification in land use planning makes this conflicting relationship even more pronounced because densification inherently deals with the already built environment. Planning therefore takes place within a tight web of existing rights and duties engraved in complex institutional norms and regulations. Potential for redevelopment is often given, but the land is frequently not accessible due to the land rights secured by strongly protected property titles. In this situation, planning often fails to deal with complex private property right arrangements as most avenues of public intervention were crafted to handle simpler property rights situations on unbuilt agricultural or former industrial land. However, in a context of land scarcity, land use planning needs to cope with complex property rights situations on already built land such as intermixed parcels of different sizes, co-ownership constellations, rights to object granted to neighbors, rights of way, mosaics of easements, etc. (for discussion, e.g. Blomley, 2008, 2017; Gerber et al., 2018). More than ever, therefore, a keen understanding of the close interactions between public policy and property rights is required to effectively steer (socially) sustainable spatial development.

### 2.2.5 *Actors' Use Strategies*

Ultimately, the IRR postulates that resources development is not only influenced by formal institutions. Moreover, actors and their appropriation strategies also play a significant role in this complex process of negotiation (Healey, 2007a, 2007b). Even though formal rules (legally) frame resource users in their activities, users can simultaneously exercise their own agency within this frame of reference in order to take advantage of the opportunities granted by those rules. In other words, formal institutions are regarded as the product of a socio-political compromise crystallizing in space and time the complex power relationships shaping resource uses. This compromise is never stable, as laws and regulations are constantly revised, remain unimplemented, and can be diverted or even hijacked by different actors involved (Gerber et al., 2020: 157). The IRR analytical framework aims to capture how these different actors use interests to evolve within a given institutional setting in order to understand how sustainable outcomes are shaped.

An actor describes "a unit that acts as a bearer of social roles with specific orientations (values, attitudes, and motivations) in a social situation. The unit of action is carried not only by individuals, but also by social structures and collectives" (Parsons, 1986 in Hillmann, 1994: 6). Hence, actors are not explicitly considered individuals. Individuals can be actors, but only if they represent interest groups or a unit with particular interests (Knoepfel et al., 2011: 60). The most common case is that actors act collectively. This includes groups of individuals who are linked by the same interests such as legal collectives (legal entities, parties, associations, unions, social groups) or social entities (e.g. an administrative unit within the city administration) (Hengstermann, 2019: 113).

On the one side, actors' interests, ideas, and values depend on the configuration of the institutions in force. For example, landowners are in a position of power due to the protection guaranteed by private property rights. They can enact their objectives in a very targeted manner. On the other side, an actor's behavior is itself goal-oriented and strategic (Hall & Taylor, 1996: 955). Within a given institutional setting, actors develop strategies to defend their own interests in order to achieve specific goals. For instance, within the structure of incentives produced by formal rules and norms, actors regularize or adapt their plans and actions to defend their own interests and objectives (Ostrom, 2007: 23). Consequently, an actor's behavior may be influenced both by reference to a familiar set of moral obligations and by strategic calculation about what others will do. This brings to the forefront that institutions themselves are never stable or predetermined. On the contrary, they are the result of social practice and construction, strategic actors' behavior, and decision-making (for discussion, e.g. Drahos, 2004; Fuys & Dohrn, 2010; Gerber et al., 2018; Hess & Ostrom, 2003) (see this chapter). Actors can either (re)activate, change, or (re)formulate existing formal rules through targeted intervention strategies in order to defend their own interests effectively.

Another way to understand the relationship between actors, institutions, and strategies is the "game-actors-play" (Scharpf, 1997) metaphor.

Players (actors) play on a joint playground (arena) and move within determined rules of the game (institutions). To understand the game, its processes and outcomes, players must not only understand the rules of the game, but also the strategies of the other players. However, these strategies are not made explicit but can be read and understood through each move and decision they make. To win, players must mobilize available "policy resources" such as capital, personnel, infrastructure, information or know-how.[4] Policy resources are understood as means "actors use to assert their values and interests in different stages of the process" (Knoepfel et al., 2011: 86). The players can combine these policy resources depending on their availability and strategic background considerations. Simultaneously, they must also invest in the creation and maintenance of these policy resources to maintain or increase their availability in the long term. A rich equipment and skillful combination of policy resources by the players points to influential or powerful actors who, in principle, have a great chance to win the game (for discussion, Hengstermann, 2019: 95–108).

In general, the IRR distinguishes between regulators, owner-, and user actors each of which can guide, structure, or even determine housing use through their strategic behavior. In Chapter 3.5, the strategies employed by the actors involved in the decision-making process about residential densification are outlined in more detail.

## References


Aalbers, M. B. (2017). The variegated financialization of housing. *International Journal of Urban and Regional Research, 41*, 542–554.

Aalbers, M. B. (2019). Financialization. In D. Richardson, N. Castree, M. F. Goodchild, A. L. Kobayashi, & R. Marston (Eds.), *The international encyclopedia of geography. People, the earth, environment, and technology*. Wiley.

Aubin, D. (2007). *L'eau en partage: Activation des règles dans les rivalités d'usages en Belgique et en Suisse*. P.I.E. Peter Lang.

Balmer, I., & Gerber, J.-D. (2017). Why are housing cooperatives successful? Insights from Swiss affordable housing policy. *Housing Studies, 33*, 361–385.

Becker, G. S. (1976). *The economic approach to human behavior*. The University of Chicago Press.


[4] Knoepfel and Nahrath 2007: 184) distinguishes between ten policy resources actors can activate: information, law, personnel, organization, consensus, infrastructure, political support, time, money, and violence.

Benjaminsen, T., & Svarstad, H. (2019). Political ecology. *Encyclopedia of Ecology, 4*(2), 391–396.
Berger, P. L., & Luckmann, T. (1966). *The Social construction of reality: A treatise in the sociology of knowledge*. Doubleday.
Blomley, N. (2008). Making private property: Enclosure, common right and the work of hedges. *Social & Legal Studies, 17*(03), 311–331.
Blomley, N. (2017). Land use, planning, and the difficult character of property. *Planning Theory & Practise, 18*(3), 351–364, https://doi.org/10.1080/14649357.2016.1179336.
Blomquist, W. (2012). A political analysis of property rights. In D. Cole, & E. Ostrom (Eds.), *Property in land and other resources* (pp. 369–384). Lincoln Institute.
Bookchin, M. (1993). What is social ecology? In M. Zimmermann, M. Callicott, J. B. Clark, K. J. Warren, & I. J. Klavier (Eds.), *Environmental philosophy: From animal rights to radical ecology* (pp. 45–78). AK Press.
Bourdieu, P. (1977). *Outline of a theory of practice*. Cambridge University Press.
Bourdieu, P. (1998). *Practical reason*. Stanford University Press.
Bromley, D. W. (1991). *Environment and economy: Property rights and public policy*. Blackwell.
Bromley, D. W. (1992). The commons, common property, and environmental policy. *Environmental and Resource Economics, 2*, 1–17.
Coase, R. H. (1960). The problem of social cost. *The Journal of Law & Economics, 56*(4), 837–877.
Commons, J. R. (1893). *The distribution of wealth*. Macmillan.
Constant, B. (1988). *Political writings*. Cambridge University Press.
Cooter, R., & Ulen, T. (2004). *Law and economics*. Addison-Wesley.
de Buren, G. (2015). *Understanding resource management: An introduction into institutional resource regime (IRR) and a field guide for empirical analysis*. sanu durabilitas & idheap Lausanne.
Dembski, S., & Salet, W. (2010). The transformative potential of institutions: How symbolic markers can institute new social meaning in changing cities. *Environment and Planning A, 42*, 611–625.
Demsetz, H. (1967). Towards a theory of property rights. *The American Economic Review, 57*(2), 347–359.
Drahos, P. (2004). The regulation of public goods. *Journal of International Economic Law, 7*(2), 321–339.
Ehrlich, P. (1986). *The population bomb*. Ballantine Books.
Engels, F. (1872). *Zur Wohnungsfrage*. Marx-Engels Werke (MEW), Band 18.
Ensminger, J. (1992). Making a market—The institutional transformation of an African Society. In J. Ensminger (Ed.), *Political economy of institutions and decisions*. Cambridge Press.

Enzensberger, H. M. (1974). A critiue of political ecology. *New Left Review, 8*, 3–32.

Evans, B., Joas, M., Sundback, S., & Theobald, K. (2006). Governing local sustainability. *Journal of Environmental Planning and Management, 49*(6), 849–867.

Evans, J. P., & Jones, P. (2008). Rethinking sustainable urban regeneration: Ambiguity, creativity, and the shared territory. *Environment and Planning A, 40*(6), 1416–1434.

Fuys, A., & Dohrn, S. (2010). Common property regimes: Taking a closer look at resource access, authorization, and legitimacy. In L. German, J. J. Ramisch, & R. Verma (Eds.), *Beyond the biophysical: Knowledge, culture, and politics in agriculture and natural resource management* (pp. 193–214). Springer.

Gerber, J.-D., & Debrunner, G. (2021). Power and planning. A theory-informed discussion of a large housing development project in Zurich, Switzerland. *Land Use Policy.* https://doi.org/10.1016/j.landusepol.2022.106400

Gerber, J.-D., & Nahrath, S. (2013). *Beitrag zur Entwicklung eines Ressourcenansatzes der Nachhaltigkeit*. University Press.

Gerber, J.-D., Hartmann, T., & Hengstermann, A. (2018). *Instruments of land policy—Dealing with scarcity of land* (J.-D. Gerber, T. Hartmann, & A. Hengstermann, Eds.). Routledge.

Gerber, J.-D., Koepfel, P., Nahrath, S., & Varone, F. (2009). Institutional resource regimes: Towards sustainability through the combination of property-rights theory and policy analysis. *Ecological Economics, 68*(3), 798–809.

Gerber, J.-D., Lieberherr, E., & Knoepfel, P. (2020). Governing contemporary commons: The Institutional Resource Regime in dialogue with other policy frameworks. *Environmental Science and Policy, 112*(6), 155–163.

Gerber, J.-D., Nahrath, S., & Hartmann, T. (2017). The strategic use of time-limited property rights in land-use planning: Evidence from Switzerland. *Environment and Planning A, 49*(7), 1684–1703.

Hagmann, T., & Hoehne, M. V. (2009). Failues of the state failure debate: Evidence from the Somali territories. *Journal of International Development, 21*, 42–57.

Hall, P., & Taylor, R. (1996). Political science and the three new institutionalisms. *Political Studies, 44*, 936–957.

Haller, T. (2019). Towards a new institutional political ecology: How to marry external effects, institutional change and the role of power and ideology in commons studies. In T. Haller, T. Breu, T. De Moor, C. Rohr, H. Znoj (Eds.), *The commons in a glocal world: Global connections and local responses. Earthscan studies in natural resource management.* Routledge.

Haller, T., Acciaioli, G., & Rist, S. (2016). Constitutionality: Conditions for crafting local ownership of institution-building processes. *Society and Natural Resources, 29*(1), 68–87.
Hallowell, A. I. (1943). The nature and function of property as a social institution. *Journal of Legal and Political Sociology, 1*, 115–138.
Hardin, G. (1968). The tragedy of the commons. *Science, 162*(3859), 1243–1248.
Hardin, G. (1991). The tragedy of the unmanaged commons: Population and the disguises of providence. *Trends in Ecology & Evolution, 9*(5), 175–199.
Harloe, M. (1982). Towards the decommodification of housing? A comment on council house sales. *Critical Social Policy, 2*(4), 39–42.
Hartmann, T., & Needham, B. (2012). *Planning by law and property rights reconsidered* (T. Hartmann & B. Needham, Eds.). Ashgate.
Harvey, D. (1985). *The urbanization of capital: Studies in the history and theory of capitalist urbanization.* Johns Hopkins University Press.
Harvey, D. (2005). *A brief history of neoliberalism.* Oxford University Press.
Harvey, D. (2012). *Rebel cities: From the right to the city to the urban revolution.* Verso.
Healey, P. (1996). The communicative turn in planning theory and its implications for spatial strategy formation. *Environment and Planning b: Planning and Design, 23*(2), 217–234.
Healey, P. (1999). Institutionalist analysis, communicative planning, and shaping places. *Journal of Planning Education and Research, 19*(2), 111–121.
Healey, P. (2007a). The new institutionalism and the transformative goals of planning. In Verma, N. (Ed.), *Institutions and planning.*
Healey, P. (2007b). *Urban complexity and spatial strategies: Towards a relational planning for our times.* Routledge.
Hengstermann, A. (2019). *Von passiver Bodennutzungsplanung zu aktiver Bodenpolitik—Eine Fallstudie zur Wirksamkeit von bodenpolitischen Instrumenten anhand von Lebensmittel-Discountern* (1st ed.). Springer VS.
Hess, C., & Ostrom, E. (2003). Ideas, artifacts, and facilities: Information as a common-pool resource. *Law and Contemporary Problems, 66*, 111–145.
Hillmann, K.-H. (1994). *Wörterbuch der Soziologie* (1st ed.). Kröner.
Hobbes, T. (1651). *Leviathan or the matter, forme and power of a commonwealth.* Basil Blackwell.
Hodkinson, S. (2012). The return of the housing question. *Ephemera: Theory & Politics in Organization, 12*(4), 423–444.
Immergut, E. M. (1998). The theoretical core of the new institutionalism. *Politics & Society, 26*(1), 5–34.
Jacobs, H. M. (1998). *Who owns America?* University of Wisconsin Press.

Jacobs, H. M., & Paulsen, K. (2009). Property rights: The neglected theme of 20th-century American planning. *Journal of the American Planning Association, 75*(2), 134–143.

Keynes, J. M. (1936). *The general theory of employment, interest, and money*. Macmillan.

Knoepfel, P. (1986). Distributional issues in regulatory policy implementation—The case of air quality control policies. In A. Schnaiberg, N. Watts, & K. Zimmermann (Eds.), *Distributional conflicts in environmental policy* (pp. 363–379). Gower.

Knoepfel, P., & Nahrath, S. (2007). Environmental and spatial development policy. In U. Klöti, P. Knoepfel, & H. Kriesi (Eds.), *Handbook of Swiss politics* (pp. 705–733). NZZ Verlag.

Knoepfel, P., Csikos, P., Gerber, J.-D., & Narath, S. (2012). Transformation der Rolle des Staates und der Grundeigentümer in städtischen Raumentwicklungsprozessen im Lichte der nachhaltigen Entwicklung. *Nomos Verlagsgesellschaft, 53*(3), 414–443.

Knoepfel, P., Kissling-Näf, I., & Varone, F. (2001). *Institutionelle Regime für natürliche Ressourcen: Boden, Wasser und Wald im Vergleich*. Helbing & Lichtenhahn.

Knoepfel, P., Kissling-Näf, I., & Varone, F. (2003). *Institutionelle Regime natürlicher Ressourcen in Aktion*. Helbing & Lichtenhahn.

Knoepfel, P., Larrue, C., Varone, F., & Hill, M. (2011). *Public policy analysis*. Policy Press.

Koelble, T. A. (1995). The new institutionalism in political science and sociology. *Comparative Politics, 27*(2), 231–243.

Krueger, R., & Agyeman, J. (2005). Sustainability schizophrenia or "actually existing sustainabilities?" toward a broader understanding of the politics and promise of local sustainability in the US. *Geoforum, 36*(4), 410–417.

Lasswell, H. D. (1936). *Politics: Who gets what, when, how* (1st ed.). Springer.

Locke, T. (1689). *Labor theory of ownership*. In Second Treatise of Government. Walter Euchner 1977.

Mackay, F., Kenny, M., & Chappell, L. (2010). New Institutionalism through a gender lens: Towards a feminist institutionalism? *International Political Science Review, 31*(5), 573588.

Mahoney, J., & Thelen, K. (Eds.). (2010). *Explaining institutional change: Ambiquity, agency and power*. Cambridge University Press.

Malthus, T. (1789). *Essay on the principle of population*. Printed for J. Johnson in St. Paul's Church-Yard.

Mandelbaum, S. J. (1985). The institutional focus of planning theory. *Journal of Planning Education and Research, 5*, 3–9.

March, J. G., & Olsen, J. P. (1989). *Rediscovering institutions: The organizational basis of politics*. Free Press.

Marcuse, P. (2012). A critical approach to solving the housing problem. In N. Brenner, P. Marcuse, & M. Mayer (Eds.), *Cities for people, not for profit*. Routledge.

Martínez-Alier, J. (2002). *The environmentalism of the poor: A study of ecological conflicts and valuation*. Edward Elgar.

Marx, K. (1859). *Zur Kritik der politischen Ökonomie*. Karl Marx/Friedrich Engels—Werke. (Karl) Dietz Verlag.

Marx, K. (1868). *Über die Nationalisierung des Grund und Bodens*. Karl Marx/ Friedrich Engels—Werke. (Karl) Dietz Verlag.

Mayntz, R., & Scharpf, F. W. (1995). Der Ansatz des akteurszentrierten Institutionalismus. In R. Mayntz & F. W. Scharpf (Eds.), *Gesellschaftliche Selbstregelung und politische Steuerung* (pp. 39–72). Campus Verlag.

McCarthy, J. (2002). First world political ecology: Lessons from the wise use movement. *Environment and Planning A, 34*(7), 1281–1302.

Meyer, J. M. (2009). The concept of private property and the limits of the environmental imagination. *Political Theory, 37*(1), 99–127.

Miller, D. (1978). Democracy and social justice. *British Journal of Political Science, 8*(1), 1–19.

Needham, B., & Verhage, R. (1998). The effects of land policy: Quantity as well as quality is important. *Urban Studies, 35*(1), 25–44.

Neuman, M. (2005). The compact city fallacy. *Journal of Planning Education and Research, 25*(11), 11–26.

Nicol, L. A. (2011). The role of institutional regimes in motivating change for sustainable housing. *Building Research and Information, 39*(5), 459–472.

Nicol, L. A., & Knoepfel, P. (2008). Institutional regimes for sustainable collective housing stocks. *Swiss Political Science Review, 14*(1), 157–180.

North, D. C. (1990). *Institutions, institutional change, and economic performance*. University Press.

North, D. C. (1991). Towards a theory of institutional change. *Quarterly Review of Economics and Business, 31*(4), 3–12.

North, D. C. (1994). Economic performance through time. *The American Economic Review, 84*(3), 359–368.

Ostrom, E. (1990). *Governing the commons: The evolution of institutions for collective action*. Cambridge University Press.

Ostrom, E. (1992). Governing the commons: The evolution of institutions for collective action. *Natural Resources Journal, 32*(2), 415–430.

Ostrom, E. (2005). *Understanding institutional diversity, public choice*. Princeton University Press.

Ostrom, E. (2007). Institutional rational choice: An assessment of the institutional analysis and development framework. In P. Sabatier (Ed.), *Theories of the policy process* (pp. 21–64). Westview Press.

Ostrom, E. (2009). A general framework for analyzing sustainability of social-ecological systems. *Science, 325*(5939), 419–422.
Powell, W., & DiMaggio, P. (1991). *The new institutionalism in organizational analysis*. University of Chicago Press.
Proudhon, P.-J. (1840) *Was ist das Eigentum? erste Denkschrift*. Verlag Monte Verita. Reprint 1971.
Ribot, J. C., & Peluso, N. L. (2003). A theory of access. *Rural Sociology, 68*(2), 153–181.
Ricardo, D. (1812). *The principles of political economy and taxation*. Penguin Books.
Robbins, P. (2004). *Political ecology: A critical introduction*. Blackwell Publishing.
Rolnik, R. (2013). Late neoliberalism: The financialization of homeownership and housing rights. *International Journal of Urban and Regional Research, 37*(3), 1058–1066.
Savini, F., Majoor, S., & Salet, W. (2015). Dilemmas of planning: Intervention, regulation, and investment. *Planning Theory, 14*(3), 296–315.
Sax, J. L. (1992). The constitutional dimensions of property: A debate. *Loyola of Los Angeles Law Review, 26*(3), 23–37.
Scharpf, F. W. (1997). *Games real actors play: Actor-centered institutionalism in policy research*. Westview Press.
Scharpf, F. W. (2000). *Interaktionsformen. Akteurzentrierter Institutionalismus in der Politikforschung*. VS Verlag für Sozialwissenschaften.
Schipper, S. (2014). The financial crisis and the hegemony of urban neoliberalism: Lessons from Frankfurt am main. *International Journal of Urban and Regional Research, 38*(1), 236–255.
Schlager, E., & Ostrom, E. (1992). Property-rights regimes and natural resources : A conceptual analysis. *Land Economics, 68*(3), 249–262.
Schumacher, E. F. (1973). *Small is beautiful*. Harper and Row.
Shahab, S., Clinch, J. P., & O'Neill, E. (2019). Impact-based planning evaluation: Advancing normative criteria for policy analysis. *Environment and Planning b: Urban Analytics and City Science, 46*(3), 534–550.
Smith, A. (1776). *An inquiry into the nature and causes of the wealth of nations*.
Steiger, O. (2006). Property economics versus new institutional economics: Alternative foundations of how to trigger economic development. *Journal of Economic Issues, 40*(1), 183–208.
Svarstad, H., Benjaminsen, T. A., & Overå, R. (2018). Power theories in political ecology. *Journal of Political Ecology, 25*(1), 350–363.
Swyngedouw, E. (2009). The political economy and political ecology of the hydro-social cycle? *Journal of Contemporary Water Research and Education, 142*(1), 56–60.

Thelen, K. (2003). How institutions evolve: Insights from comparative historical analysis. In J. Mahoney & D. Rueschenmeyer (Eds.), *Comparative historical analysis in the social sciences* (pp. 208–240). Cambridge University Press.

Thomann, E., Hupe, P., & Sager, F. (2018). Serving many masters: Public accountability in private policy implementation. *Governance, 31*(2), 299–319.

Vatn, A. (2005). *Institutions and the environment*. Edward Elgar Publishing.

Viallon, F.-X. (2018). Added value capturing in Switzerland: How much is enough? In J.-D. Gerber, T. Hartmann, & A. Hengstermann (Eds.), *Instruments of land policy—Dealing with scarcity of land* (pp. 59–68). Routledge.

Williamson, O. E. (2000). The new institutional economics: Taking stock, looking ahead. *Journal of Economic Literature, 38*(3), 595–613.

Wimark, T., Andersson, E. K., & Malmberg, B. (2019). Tenure type landscapes and housing market change: A geographyical perspective on neoliberalization in Sweden. *Housing Studies*, 1–24.

Zimmerer, K. S., & Bassett, T. J. (Eds.). (2003). *Political ecology: An integrative approach to geography and environment-development studies*. Guildford Press.

CHAPTER 3

# The IRR Applied to Housing: Governing Densification for Socially Sustainable Housing Development

## 3.1 Housing Study Focus: Governing Densification for Socially Sustainable Housing Development

In the following chapters, the theoretical approach and the IRR analytical framework introduced (Sects. 2.5–2.9) are used to understand the implementation process of densification goals in the urban housing segment (study focus of this book). More precisely, the actor-centered new institutionalist political ecology approach (Chapter 2) enables researchers to analyze and to explain why many different stakeholders can come into resource conflict with each other in a densifying urban environment. It makes possible to trace the concrete mechanisms of (unfair) power distribution and the socio-political structures involved that help to understand why some actors tend to win while others lose. By socio-political it is meant that legal configurations (laws and policies) (Sect. 3.5) are analyzed as well as actors' decision-making behavior (Sect. 3.6).

## 3.2 Governance of Densification for Sustainable Urban Housing Development

Governance in this book is generally appraised as "the interactions among structures, processes and traditions that determine how power and responsibilities are exercised, how decisions are taken, and how citizens or



G. Debrunner, *The Business of Densification*,
https://doi.org/10.1007/978-3-031-49014-9_3

other stakeholders have their say" (Graham et al., 2003: 2). It is about spheres of public debate, partnership, interaction, dialogue, and indeed conflict and dispute entered into by citizens, civil organizations, and by public authorities (Evans et al., 2006: 850).

Making explicit the governance mechanisms of possible sustainability trade-offs and power games among actors involved in densification procedures is a new contribution of this book to international land use planning, and urban and housing research. Neglecting the governance mechanisms at play in which actors succeed or fail in defending their interests—especially regarding their influence on decision-making procedures leading to social exclusion, segregation, or inequality—enable to make power structures among actors explicit. By answering the question *how* and *why* various actors involved in densification procedures question, disrupt, modify, and use the socio-political setting to appropriate resources, and thus potentially cause them to change, this research project brings new insights to the mentioned disciplines.

Figure 3.1 summarizes the three key elements on which the governance understanding (see first paragraph, Sect. 3.2) of the IRR analytical framework is built on: institutions including formal policy instruments (independent variable), actors' use strategies (intermediate variable), and the condition of the housing resource (dependent variable). By analyzing the governance mechanisms at play (black arrows in Fig. 3.1) between the three main variables, the IRR analytical approach enables to explain why some groups or interests experience disproportional access to housing and tend to lose while others tend to win (power relations). In other words, the identified governance mechanisms are defined as places of power where actors involved in densification procedures are able to influence the "rules of the game" (North, 1994), their activation or implementation process in a targeted way (Gerber & Debrunner, 2022). Depending on the institutional setting, stakeholders have different means to either change these intervention ways or try to influence others to do so. These mechanisms are at the same time the channels through which stakeholders exercise power.

In the following chapter, I will explain each row, column, and arrow of Fig. 3.1, which explains the governance of densification procedures, in greater detail. I start with the dependent variable—the condition of the housing stock resource—and proceed with the independent (Sect. 3.5) and intermediary (Sect. 3.6) variables.

| Independent variable: **Institutions** | **Policy instruments** | Intermediary variable: **Users of the resource** | Dependent variable: **Condition of the housing stock resource** |
|---|---|---|---|
| **Political system (e.g. federalism)**<br>↓↑<br>**Institutional resource regime (IRR):**<br>Public policies (legitimize state intervention)<br>- Housing law<br>- Planning law<br>- Building law<br>- Environmental law<br>- Monument conservation<br>- Tax law<br>↓↑<br>Property rights (protect private interests against state intervention)<br>- Property rights<br>- Tenancy law<br>- Contract law<br>- Mortgage law<br>↓↑<br>**Ideology, social norms & values** | ⇄ **Public policies:**<br>(1) Subsidies<br>(2) Zoning<br><br>**Property rights:**<br>(3) Contracts<br>(4) Ownership | ⇄ **Planning administrations:**<br>- Municipality<br>- Canton/region<br>- National<br><br>**Owner actors:**<br>- Property owners<br>- Investors<br>- Developers<br>- Banks<br>- Private real-estate and building industry<br><br>**Residents & end users:**<br>- Tenants<br>- Neighbors<br>- Non-profit organizations | ⇄ **Social sustainability in housing**<br><br>**Appraised along the indicators of e.g.,**<br>- Housing affordability, availability, accessibiity<br>- Housing quality in and around the building<br>- Community cohesion<br>- Citizenship & decision-making |

**Fig. 3.1** Governance of densification for sustainable urban housing development (overview): The IRR analytical framework applied to the study of housing stocks in dense city areas (Gerber et al., 2009, 2020; Knoepfel et al., 2007; Nicol & Knoepfel, 2008)

## 3.3 Housing Stock: Evaluating the Condition of a Resource

Although the IRR has traditionally been applied to the study of natural resources (e.g. land, water), the framework is also well suited for the analysis of non-natural resources. Because analytically, "neither the environment as such or parts of features of the environment per se are resources; they become resources only if, when, and in so far as they are, or are considered to be, capable of serving man's needs. The word 'resource' is an expression of appraisal and, hence, a purely subjective concept" (Zimmermann, 1933: 3). The IRR identifies the definition of resources as socially constructed. A resource is acknowledged as inseparable from humans and their wants. It is defined by its use or its integration into an economic or political context (Kébir, 2010: 70). The use of a resource is dependent on its use value that results from the respective social context within which human goals and capabilities are shaped (Bathelt & Glückler, 2005: 1547).

By following a resource-oriented approach, the IRR analytical approach enables to take multiple and conflicting resources use situations into consideration (see Sect. 2.5). Each resource (be it housing or others such as water or green spaces) creates unique use(r) constellations in which various actors find themselves as rivals. The IRR thus follows a more comprehensive approach as it is capable of portraying the complexity of heterogenous use situations. This becomes all the more relevant in dense urban environments since resources such as urban land get scarce and contested, and various actors request use or disposal rights upon them.

In this book, different goods and services produced by the resource "housing stock"—the use of which is significantly changed by densification—are analyzed in detail. While "housing" is generally used to describe the construction and usage of buildings in which people live in, the term "housing stock" refers to the total number of dwellings (houses, flats, maisonettes, bed-sits etc.) in a given area (Balmer & Bernet, 2015). Indeed, housing stocks as a resource may be used for different purposes: for shelter *but also* for non-residential services such as investment, energy supply, urban design, or immigration. Energy suppliers, for example, rely on the demand created by housing stocks to sell their product. Urban designers rely on the physical characteristics of the buildings to create suitable urban space (Nicol, 2011: 459). Moreover, the housing stock is a resource that is unique in a few respects. It is a durable and long-lasting resource often existing for more than a century (Balmer & Bernet, 2015: 181). It is also an economically significant resource. Traded on the free market, housing is a commodity with enormous economic potential which is why it is often treated as a highly valued collateral (e.g. Aalbers, 2017).

Especially in cities, where demand for housing is high and the potential for capital accumulation is lucrative, the competition between actors interested in using urban land for housing investment is rising and rents increase simultaneously. The obligation to promote densification introduced by many states and cities has even reinforced this competition leading to scarcity of space and corresponding land and rent value increases (Breheny, 1992: 143; Harvey, 2012: 127). This financial potential of urban housing stocks is, however, in several ways juxtaposed with the role of housing to provide the basic necessity of shelter. According to the Universal Declaration of Human Rights[1] (Article 25), housing represents a basic human need and essential good. When residents do not have the funds necessary to access housing provided by the free market then the provision of the basic human right for shelter is foregone un-less governmental action is taken (Brenner et al., 2012: 224; Schönig, 2020; Schönig et al., 2017).

Finally, the housing stock in many cities—due to increased $CO_2$ emissions in old buildings—presents a target for energy policy efforts (see e.g. UN Agenda 21; UN Habitat Agenda). Through the renovation and conversion of existing apartments, the energy requirement per capita is to be reduced (e.g. Bhatti, 2001; Bhatti & Dixon, 2003; Næss & Saglie, 2019; Priemus, 2005). To limit energy emissions, upgrading of existing, partially historically protected buildings is absolutely necessary, which in turn has a direct influence on rent prices after the reconstruction and the preservation of the architectural heritage (Nicol, 2013). In sum, housing has a crucial role to play in the sustainable development of cities due to its various functions for different groups and individuals.

The decisions taken by actors involved in specific housing uses must be considered in detail if the question of housing sustainability is to be thoroughly addressed (Nicol, 2011: 459). This objective can only be attained, however, if all users jointly ensure that the quantities they extract or withdraw from the stock do not reach the limit of the reproductive capacity of the resource system (Gerber et al., 2009: 800). Otherwise, rivalries between different user actors occur because the use of a good or service extracted from the housing resource interferes with the use of other goods and services by another actor (Nicol & Knoepfel, 2008: 161). In the following section, the aim is to emphasize what a (socially) sustainable status of housing conditions might look like so that user conflicts do not lead to overuse or the depletion of the housing resource.

[1] UN General Assembly (1948): Universal Declaration of Human Rights, Paris.

## 3.4 Housing Sustainability: With a Focus on Its Social Dimension (Dependent Variable)

Even though the IRR analytical framework has always been closely linked to the political and scientific debates about sustainability, evaluating the concept in regard to the interactions between its three dimensions—ecological, economic, *and* social—has largely been missing so far (see e.g. Nicol, 2013; Hengstermann, 2019). Mainly because the IRR has an often-unstated bias towards ecological sustainability (reproduction of the capital) as it evolved out of environmental debates of the 1980s (Gerber et al., 2020). Its original focus built on a "hard or strong sustainability" (Jacobs, 1992) perspective which implies that renewable resources must not be drawn down faster than they can be renewed. Natural capital must not be spent, but we must live off the income produced by the capital.

Sustainability, in other words, was understood as the physical end-state of the environment that can be sustained over time, while sustainable development was seen as a process of change towards achieving sustainability goals (Marcuse, 1998). The primary attempt of the IRR approach was to regulate the emission of pollutants in a way that leads to less environmental degradation. The presumed neutrality of ecology as a science when entering environmental debates was moreover seen as illusory (Knoepfel et al., 2001, 2007) because political choices were identified as responsible for environmental impacts and resources degradation. Hence, even though the evolvement of the IRR approach has always been strongly linked to Neo-Marxist political ecology perspectives (see Sect. 2.4, e.g. Robbins, 2004), *and* the study of power relations in political decision-making. The connection between power *and* sustainability has not been made explicit since power was regarded as an integral element of the rules in force (e.g. property rights).

Therefore, in light of this book, greater emphasis is put on the relationship between power *and* sustainability to understand which factors provide certain individuals with the power to defend their own interests in resource use. Furthermore, to understand this complexity, greater awareness is being put on the social dimension of sustainability that allows for capturing the origins, causes, and effects of socio-environmental disputes in more detail. This is done by the structured analysis of the actors' constellations, their resource use strategies (of owners and non-owners), and their decision-making behavior within the agency.

In particular, my intention in the next chapters is to advance a conceptualization of social sustainability in regard to housing challenges in dense city areas (Sect. 3.4.2) in order to "bring light into the dark" and to make a rather fuzzy concept feasible and transparent for my research. Stressing the unique social features of sustainable housing development presents my first step towards an interpretation that is sufficiently rigorous to provide useful tools for practical analysis and effective policy-making.

### *3.4.1 Social Sustainability as an Analytical Framework of Reference*

Despite the nearly universal acknowledgment that sustainable cities are a desirable policy goal (for discussion, e.g. Barbier, 1987; Khan, 1995), there is still less certainty about what this might mean in practice and how to define the concept in any analytically rigorous way (Williams, 2010). Since the popular release of the Brundtland report in 1987[2] (WCED, 1987), many deductive approaches have been published on the conceptualization of sustainability (for discussion, e.g. Christen & Schmidt, 2012; Elliot, 1999; Jabareen, 2008; Redclift, 2005).

In the late 1980s and early 1990s, the concept of sustainability has primarily been understood to endorse the pursuit of economic growth under the condition that the environment will not be damaged too much. The goal was to seek maximum economic growth while minimizing environmental degradation (e.g. Cernea, 1993). The driving force behind the conception of sustainability was "the belief that if the environment continues to be degraded, economic growth will be stifled. Therefore, economic growth can only be sustained if attention is paid to sustaining resources and the environment" (Portney, 1994: 830). In a second variation, academic literature mainly focused on the environment. A sustainable city was associated with efficient waste management,

[2] In 1987, the World Commission on Environment and Development published Our Common Future, also known as the Brundtland Report. So far, the commission has released the most popular definition of "sustainable development" by conceptualizing it as "development that meets the needs of the present withoutcompromising the ability of future generations to meet their own needs" (WCED, 1987: 43). Ever since, the meaning of the concept has been debated passionately. Even though the definition is no doubt,very clever, it is not necessarily the clearest. For instance, it does not adequately articulate the disaggregated branches of different functions of society and biosphere in a manner which, on the onehand, demonstrates their separate mechanisms and, on the other, reveals their integrating roles to constitute what is being termed sustainable development (Khan, 1995).

recycling opportunities, reduced car dependency and greater use of alternative modes of transport in order to limit cities' ecological footprint (e.g. Bromley et al., 2005). The primary aim was to sustain the physical environment while economic growth was considered of secondary importance.

In *housing*, for instance, the energy crises of the 1970s created awareness for building environmentally sustainably. Ideas of energy efficiency and concern about the wasteful use of fossil fuels became increasingly promoted. Technological advancements with renewable materials, energies, and construction techniques were introduced that helped to reduce the destructive environmental impacts of housing (Nicol & Knoepfel, 2008: 158). The main intention was to prevent environmental damage by the observation that every action taken on behalf of economic growth should seek to be as environmentally sensitive as possible (Portney, 1994: 830). Later on, this idea strongly influenced the development of the IRR approach, which evolved out of the environmental debates of the 1980s. It intended to deal with the use of the environment as a sink for pollution, therefore attempting to regulate the emission of pollutants, but with considering resources use capacity as a whole (Gerber et al., 2020).

As this summary of the literature shows, however, the social dimension of sustainability has only received little attention in policy, planning, and academia so far even though it has been more than forty years since the Brundtland report's release (for discussion, e.g. Foladori, 2005; Manzi, 2010; Murphy, 2012; Weingaertner & Moberg, 2014; Woodcraft, 2012). Within the densification debate, for example, for many years the least explored and most ambiguous claim was that the compact city is socially equitable (Burton, 2003: 538). During the 1980s, urban regeneration projects focused mainly on economic and environmental aspects of degraded inner-city areas while neglecting social aspects (Rérat, 2012: 116). Nevertheless, emphasizing the environmental and economic spheres exclusively has been increasingly questioned during the 1990s (for discussion, e.g. Basiago, 1999; Budd et al., 2008; Crabtree, 2006; Mitlin & Satterthwaite, 1996; Redclift, 2005). Mainly because the three sustainability dimensions were considered to be strongly interlinked. The diminishing of one affects that of the others.

Until today, however, there is still no broad consensus on the meaning of the term "social sustainability". The concept is far more difficult to quantify than economic growth or environmental impact (for discussion, e.g. IFHP, 2019; Oxford Institute for Sustainable Development, 2007). It is considered a vague and fuzzy concept that comes with a number of ethical, political, and methodological challenges (Davidson, 2010; Woodcraft, 2012). Moreover, the concept of social sustainability itself is culture-dependent (Chiu, 2003) and inherently normative in both the theoretical conceptualization and the pursuit of translating subsequent new insights into empirical research and effective policy agendas (Jonkman, 2019). A brief review of the literature shows how social sustainability has been conceptualized over time (Table 3.1).

A clear distinction must be drawn between the three concepts of "social sustainability", "social justice", and "social inclusion": while the concept of "social sustainability" has evolved out of environmental debates and is strongly related to ecological issues such as land use planning (e.g. Campbell, 1996; Elkin et al., 1991), the concept of "social justice" has mostly been addressed by philosophy (e.g. Polanyi, 1957; Rousseau, 1762) and critical social sciences (e.g. Fainstein, 2001, 2010; Harvey, 1973; Rawls, 1972; Sen, 1999). In justice literature, the question of the ethical foundation of society has challenged mankind since the dawn of civilization. The idea of social justice has become a key element of these reflections (Heidenreich, 2011). Consequently, the concept of social justice is closely linked with broader issues of social relevance such as migration or gender research resulting in insights that are extendable to a perspective on society as a whole, independently of individual preferences (Ketschau, 2015). Justice incorporates a set of topics that could best be described as attempts to improve or maintain "the quality of life of people" (Weingaertner & Moberg, 2014: 129) as well as social well-being and freedom (Woodcraft, 2012: 31). It constitutes an inherent part of sustainable development (Langhelle, 2000: 296).

In contrast, "social sustainability" inherits both a perspective on the individual and on the society (Elkington, 1999). It places explicit value on the intergenerational stability of communities (Ketschau, 2015) and on the initial control over natural resources (Barry, 1991: 238). While some social sustainability scholars concentrate on the long-term viable setting for human interaction (Biart, 2002; Yiftachel & Hedgcock, 1993), others emphasize context-sensitivity (Vallance et al., 2011) and community stability (Colantonio & Dixon, 2009; Griessler & Littig, 2005; Karlsson, 2018). Other researchers focus more on pragmatic relative improvements in resource distribution (Chiu, 2003; Polese & Stren, 2000; Portney,

**Table 3.1** Conceptualizations of social sustainability

| | |
|---|---|
| "The continuing ability of a city to function as a long-term viable setting for human interaction, communication and cultural development. [It is] about the long-term survival of a viable social unit." | Yiftachel and Hedgcock (1993: 140) |
| "[It] is the idea that our economy functions to the detriment of minorities and the poor because it forces them to disproportionately bear environmental risks." | Portney (1994: 829) |
| "A strong definition of social sustainability must rest on the basic values of equity and democracy, the latter meant as the effective appropriation of all human rights—political, civil, economic, social, and cultural—by all people." | Sachs (1999: 27) |
| "Social sustainability means development (and/or growth) that is compatible with harmonious evolution of civil society, fostering an environment conductive to the compatible cohabitation of culturally and socially diverse groups while at the same time encouraging social integration, with improvements in the quality of life for all segments of the population." | Polese and Stren (2000: 15–16) |
| "Social sustainability prescribes that an acceptable standard of housing is a necessary condition of full membership of the community (Marshall, 1964; Miller, 1978). The residents' rights are part of a general program of social advance and not necessarily expected to be fulfilled by the state at all times in every single case. More precisely, social justice in housing is interpreted as meaning that all citizens should be able to solve their housing question through voluntary transactions in the market themselves." | Bengtsson (2001: 265) |
| "Social sustainability aims to determine the minimal social requirements for longterm development (sometimes called critical social capital) and to identify the challenges to the very functioning of society in the long run." | Biart (2002: 6) |
| "To be socially sustainable, there needs to be equitable distribution and consumption of resources and assets, harmonious social relations, and acceptable quality of life." | Chiu (2003: 225) |

(continued)

**Table 3.1** (continued)

| | |
|---|---|
| "Social sustainability is given, if work within a society and the related institutional arrangements satisfy an extended set of human needs (and) are shaped in a way that nature and its reproductive capabilities are preserved over a long period of time and the normative claims of social justice, human dignity and participation are fulfilled." | Griessler and Littig (2005: 11) |
| "Social sustainability concerns how individuals, communities and societies live with each other and set out to achieve the objectives of development models which they have chosen for themselves, also taking into account the physical boundaries of their places and planet earth as a whole. At a more operational level, social sustainability stems from actions in key thematic areas, encompassing the social realm of individuals and societies, which ranges from capacity building and skills development to environmental and spatial inequalities. In this sense, social sustainability blends traditional social policy areas and principles, such as equity and health, with emerging issues concerning participation, needs, social capital, the economy, the environment, and more recently, with the notions of happiness, well being and quality of life." | Colantonio and Dixon (2009: 4) |
| "Social sustainability requires to explore how residents interpret, and incorporate concerns about, the places in which they live and the world around them." | Vallance et al. (2011: 347) |
| "People's right to be able to decide or affect their own circumstances and influence and participate in society without discrimination. This is one of the most fundamental aspects of human rights and hence of social sustainability in cities." | Karlsson (2018: 10) |

1994) (Table 3.1). In essence, the judgment, evaluation, and reflection on whether social sustainability in urban development is given or not can be made along broader principles of social (in)justice.

Thirdly, the concept of "social inclusion"—explicitly in reference to housing studies—refers to a condition, where residents live in circumstances that meet their basic needs or physical adequacy of life (Chiu,

2004; Lee & Evans, 2020; Slater, 2015). It refers to the level of integration and social position in which people are recognizably at lesser risk of experiencing a crisis, such as lack of control, residential displacement, discrimination, or loss of the home (Beer et al., 2016; Lombard, 2021). Social inclusion therefore serves as a key indicator of whether residents feel safe in their respective living situations or are confronted with the threat of losing a stable, secure, and affordable home (Cardoso et al., 2021).

Hence, the three concepts—social justice, social sustainability, *and* social inclusion—derive from different academic stands, but have always been closely related to one another. In Sects. 3.4.2 and 3.4.3, I further emphasize social sustainability and explain how the concept is assessed and operationalized in terms of housing in this book.

### *3.4.2 Social Sustainability in Housing*

For the past twenty years, much *quantitative* (see e.g. Bibby et al., 2018; Broitman & Koomen, 2015; Kyttä et al., 2013; Schmidt-Thomé et al., 2013) and *qualitative* research (see e.g. Bramley et al., 2009; Chiu, 2004; Colantonio & Dixon, 2009; Dempsey et al., 2009; Mccrea & Walters, 2012; Savini, 2011; Vallance et al., 2011) has been conducted on measuring the social impacts of urban densification. Different notions of "social sustainability" in relation to urban land use have emerged all aiming to (inter)link the stimulation of economic activities and environmental improvements with social and cultural elements of the city (Table 3.1).

To strengthen the understanding of core issues related to the concept of social sustainability in housing, indicators of how socially sustainable processes and outcomes can be evaluated are introduced (Table 3.2). The criteria figure as a well-founded and theory-based tool to operationalize and monitor the development of housing stocks (Christen & Schmidt, 2012: 405). However, as remarked in the previous section, it is important to distinguish what is actually meant by social sustainability in housing and how the term is deployed to establish relational identities and perspectives by different stakeholders (Evans & Jones, 2008: 1430). Otherwise, there is a risk that criteria based on solely individuals' norms are introduced that do not allow for a more objective assessment.

Therefore, in Table 3.2, key themes and indicators of social sustainability performance in housing are introduced that enable a systematic

**Table 3.2** Social sustainability indicators in housing. Analyzing social sustainability in densification processes takes place in close connection with ecologic and economic sustainability

| *Dimensions of social sustainability in densification projects* | | *Indicator [operational indicators]* |
|---|---|---|
| Socio-economic | **Housing Affordability**<br>Mulliner et al. (2013: 275), Stone (2006), Thalmann (2003), and Dave (2010: 13) | • **Rental costs in relation to monthly income ratio** [%]<br>• **Residual income standard** [interaction between incomes, housing costs, and non-housing necessities per person] |
| | **Housing Availability and Accessibility**<br>IFHP (2019) | • **Availability of non-profit housing** [share of non-profit housing of the total housing stock of the municipality, length of waiting lists in the non-profit housing sector]<br>• **Access criteria to housing units** [e.g. access criteria related to age, gender, income etc.] |

(continued)

**Table 3.2** (continued)

| *Dimensions of social sustainability in densification projects* | | *Indicator [operational indicators]* |
|---|---|---|
| Socio-ecologic | **Housing quality in and around the building**<br>Dave (2010: 13, 23), Alexander and Tomalty (2002: 403), and Mulliner et al. (2013) | • **Perceived satisfaction of residents with the quality of the dwelling** [perceived satisfaction with the size of home, distribution of rooms, location, rent, facilities, physical condition of the apartment and the building, daylight]<br>• **Perceived satisfaction of residents with the living quality of the local environment, namely the availability and access to facilities and amenities** [perceived satisfaction with open-, free spaces & parks; perceived availability of daily use shops and other services such as bank or post offices; leisure facilities; schools and child care; health facilities; waste and disposal options; transport nodes; employment opportunities; perceived safety during nighttime; perceived level of cleanliness; facilities for disabled people; perceived satisfaction with the urban character and the local lifestyle] |

| *Dimensions of social sustainability in densification projects* | | *Indicator [operational indicators]* |
|---|---|---|
| Social | **Community cohesion**<br>Bramley and Power (2009: 33), Dempsey et al. (2009), Dave (2010: 13), Weingaertner and Moberg (2014: 127), and Chiu (2004: 69) | • **Residential stability** [years of residency in the neighborhood]<br>• **Level of discrimination and social exclusion** [socio-economic and -demographic mix; legal protection of social mixing; perceived level of discrimination]<br>• **Community spirit and social interaction** [residents perceived level of friendliness; number of social networks and contacts (knowing people) within the neighborhood; perceived satisfaction with the social involvement in community activities and support; perceived satisfaction with the social mix of the neighborhood; perceived level of social attachment to the neighborhood] |
| | **Citizenship & Decision-making**<br>Bramley and Power (2009: 33), Dempsey et al. (2009), Dave (2010: 13), Chiu (2004: 69), and Weingaertner and Moberg (2014: 127) | • **Local democracy, participation, and empowerment** [transparency and regularity of communication between investor, municipality, and residents; protection of residents' housing needs provided by local housing, planning, and tenure security legislations] |

While each criterion may be regarded as conceptually distinct, various reinforcing relationships link them together evaluation of residential densification processes. The evaluative framework is based on a theoretical understanding of what social sustainability in housing means and is analyzed in close connection with ecologic and economic sustainability. The definition of such indicators is always subjective and depends on different normative principles as well as specific perspectives taken (Fürst & Scholles, 2001: 139; Khan, 1995: 64). However, planning for sustainable housing development does not work without the evaluation of conditions, processes, and outcomes in order to be able to assess what deficiencies might need to be addressed or what needs to remain because it corresponds with the targeted goals introduced (Budd et al., 2008: 260; Curdes, 1995: 54).

Only in doing so, planners and other practitioners are able to make a real judgment and validate the outcomes of their evaluations. This will reduce the degree of subjectivity in the policy analysis process (for discussion, e.g. Alexander & Faludi, 1989; Baer, 1997; Laurian et al., 2010; Norton, 2005; Oliveira & Pinho, 2011). Nevertheless, criteria formulation is not merely a checklist design rather it is to be seen as a necessary skill of planning professions (Shahab et al., 2019: 535). I therefore believe that the indicators introduced in Table 3.2 provide a valuable starting point for the assessment of social aspects of housing and for the detailed analysis of socially sustainable settlement transformation.

### *3.4.3 Social Sustainability in Housing as Emic Approach*

To further operationalize and to provide a comprehensive and integrated framework for analyzing social sustainability in housing, in Article 2 an emic sustainability approach is introduced. This analysis approach addresses social sustainability concerns in housing from a perspective that emphasizes community members' views on participation, negotiation, and the extent to which they can integrate their own views into the institution-building process (for discussion, e.g. Dolsak & Ostrom, 2003; Haller et al., 2016). The emic approach introduced works with the indicators presented in Table 3.2 on a broader scale but further considers the perspectives of communities and individuals' on social displacement (McCrea & Walters, 2012: 191).

The understanding *how* residents involved perceive impacts of urban (re)development on their housing livability is considered as crucial to explain whether socially sustainable development is given or not (Bramley et al., 2009: 2129). Land use challenges, such as the shift towards densification, are analyzed from the ground up as they are experienced by local communities and people actually affected by consolidation projects. The added value and legitimacy of such an emic perspective is that it is clearly most sensitive to community, local needs, and values. Any action that renders the residential community worse off is not considered compatible with urban social sustainability since the principle source of evidence is constituted by the people themselves, particularly those living in the areas in question. In other words, dense cities are not considered socially sustainable if they are not acceptable to people as places in which to live, work, and interact, or if their communities are unstable and dysfunctional (Bramley et al., 2009; Vallance et al., 2011).

This approach derives from a broad body of policy ecology literature on "actually existing sustainabilities" (for discussion, e.g. Cook & Kothari, 2001; Evans & Jones, 2008; Evans et al., 2006; Hargreaves, 2004; Krueger & Agyeman, 2005; Krueger & Gibbs, 2007). The authors of these studies claim that, regardless of *whether* development projects are obtained through intentions associated with achieving environmental justice or otherwise, they must be consistent with the goals of creating sustainable communities (Portney, 1994: 838). Stated another way, "to be socially and culturally sustainable, development must be gauged by the values that a society itself, or some member thereof, deems to be requisite for its health and welfare" (Goulet, 1971: 333). If local community members believe that change detracts from their established and preferred ways of living, they may actively resist such changes (Newman & Wyly, 2006; Wyly et al., 2010). Therefore, understanding the implications of, and reasons behind such refusals is important in two aspects (Vallance et al., 2011: 345): *first*, to effectively support bio-physical environmental goals (e.g. shift towards densification), and *second* to recognize that the pursuit of such goals might actually be counter-productive and hence unsustainable overall.

In housing, for instance, Jenks et al. (1996: 84) stated that to be truly socially sustainable, the city must have a reasonable degree of support from local residents. If not, those who can will continue to live in the city, and only the most disadvantaged will have to leave. A scenario that is highly unsustainable. The applied emic approach and the criteria used to evaluate social sustainability in housing are further introduced and conceptualized in Article 2.

## 3.5 Institutional Regime Regulating the Housing Stock (Independent Variable)

To understand how all shelter and non-shelter uses of the housing stock are governed, the IRR framework analyzes the ensemble of regulatory conditions that produce changes in the management strategies of the actors involved (Nicol, 2013). The IRR distinguishes two main categories of formalized rules—public policies *and* property rights—that operate according to a different logic and rely on opposing legitimizations.

### *3.5.1 Housing Public Policies*

Examples of public policies regulating housing under densification pressure include, for instance, housing, social welfare, environmental, or land use policies (Fig. 2.1). These are generally defined as (see e.g. Baptista & O'Sullivan, 2008; Benjaminsen & Dyb, 2008; Kemeny, 2001; Lennartz, 2011; O'Sullivan & de Decker, 2007; Ronald, 2013);

- **Housing Policy:** Housing policies encompass a range of public measures aimed at addressing housing needs, affordability, and accessibility for the population.
- **Social Welfare Policy:** Social welfare policies are designed to support individuals and families in need and ensure social equity on a general level.
- **Environmental Policy:** Environmental policies aim to address the impact of urban development on the natural environment and to promote sustainable practices.
- **Land Use Policy:** Land use policies govern how land is allocated, used, and distributed in urban areas and steer spatial development on a general level.
- **Heritage Protection or Cultural Heritage Policy:** Such policies aim to protect significant landmarks, buildings, archaeological sites, cultural landscapes, and intangible cultural heritage from damage, destruction, or inappropriate alterations.

Each of these housing public policies—regulated through public laws (see Sect. 2.6)—have a direct impact on both housing stock owners *and* other user actors such as tenants or neighbors (see Sect. 2.9). First, because public policies can place limits and restrictions on the rights of stock owners accorded them by property rights. For example, housing policies that aim to protect tenants from social exclusion may prevent an owner from raising rents according to market prices. Second, such public policies can accord use rights to persons other than the stock owner, for example, tenants or NGOs. For instance, residents' use rights can be obtained either directly from the property owner or as results of attribution or redistribution of rights resulting from the implementation of a public policy. For example, owners may be granted low-interest loans with a long payback period via tax policy, but on the condition that they build affordable dwellings (Nicol & Knoepfel, 2008: 170). Thereby, public policies can indirectly intervene in different actors' housing use interests.

### 3.5.2 *Housing Policy Instruments*

Housing policy instruments refer to specific institutional rules, laws, and regulations—enshrined in public *and* private law—aiming to shape and influence housing-related issues, such as increasing housing affordability, promoting sustainable housing development, reducing homelessness, ensuring adequate housing for all citizens, or steering tenants' or homeowners' behavior on the housing market (see Balmer & Gerber, 2017; Gerber et al., 2018). These instruments are essential legal rules used by governments, public authorities, policymakers, and other actors (e.g. homeowners) to strategically achieve specific housing outcomes or policy objectives (see Chapter 2, part II and 5.3 of this book for more details). Some common housing policy instruments include:

**Public** law housing instruments (regulated through public policy):

- **Housing financial support** (e.g. affordable housing funding programs; housing finance policies; housing subsidies or vouchers; tax incentives for developers; public housing programs; advantageous mortgages; bank guarantees; subsidized loans; homeless prevention and assistance; property tax policies; housing trust funds; housing finance regulations; land banking; housing rehabilitation programs)
- **Zoning measures** (e.g. inclusionary zoning; zones for affordable housing; zones for the protection from redevelopment; special land use zones for affordable housing provision; quotas for affordable housing; pre-emption rights; rights-to-first-refusal; added land

value capturing; housing codes and standards; density bonuses; land readjustments)

**Private** law housing instruments (regulated through the Civil Code):

- **Housing contracts** —in the context of housing often regulated through tenancy law—(e.g. rent control mechanisms; tenants' protection from redevelopment or eviction; eviction controls; urban development contracts; public–private-partnerships (PPP); affordable housing covenants; tenants' rights of appeal)
- **Housing property rights** (e.g. public land purchase; public housing; non-profit cooperative housing; expropriation; long-term ground leases on public land to non-profit housing cooperatives; community-land-trusts)

These are just a few examples of public and private law housing policy instruments that can be utilized to address specific housing challenges. The effectiveness of each instrument depends on the context and the coordinated efforts of various stakeholders involved in housing and urban development. Moreover, these housing policy instruments work in combination to create a comprehensive approach to address housing challenges and meet the diverse needs of individuals and communities. Public authorities, policymakers, but also homeowners or tenants, often strategically activate and tailor these tools to the specific circumstances and goals of their respective jurisdictions.

## 3.6 Actors' Housing Use Strategies (Intermediary Variable)

In general, the IRR distinguishes between regulators, owner-, and user actors each of which can guide, structure, or even determine housing use through their strategic behavior. In the following, the strategies employed by the actors involved in the decision-making process about residential densification are outlined.

### 3.6.1 *Active Municipal Land Policy for Sustainable Housing*

Even though there are various public actors—at the national, regional, and local level—whose decision-making directly shapes and (re)defines the housing landscape, in liberal states, the municipality is the public actor mainly responsible for dealing with housing challenges (Rudolf et al., 2018: 476). As regulators, municipal authorities (administrative level), on behalf of the city council (executive level) and the city parliament (legislative level), create, control, or dictate housing use rights.

The public regulations they apply do not directly affect the housing stock itself but the actors whose actions have direct consequences upon the stock's use and development (Nicol & Knoepfel, 2008: 166). More precisely, municipal authorities may (re)- or (de)activate different policy instruments (Article 3) to alter land parcels in size and shape and to promote tenants' social inclusion such as freehand purchase, expropriation, or pre-emption rights (e.g. Schönig, 2020; Vollmer & Kadi, 2018). The selection of these instruments is never neutral but highly political in nature. The choice corresponds to a specific interpretation of the role played by the state and/or its private partners (for discussion, e.g. Alterman, 2007; Hartmann & Spit, 2015; Needham, 2006, 2014; Needham & Verhage, 1998; Needham et al., 2018). In making this choice, municipal administrations change the use conditions for specific interest groups. For example, shifts towards owner-occupied housing may stimulate speculative activity and simultaneously undermine the availability and affordability of rental housing (Kadi & Ronald, 2014: 271). In Article 3 of this book (Sect. 5.4), I introduce and discuss four different types of intervention municipal authorities use to promote socially sustainable housing in dense cities. These intervention ways derive from the IRR approach.

Besides the legal rules and instruments, which a municipality can formally introduce to promote sustainable housing, contemporary planning scholars (e.g. Gerber et al., 2018; Hengstermann, 2019; Needham et al., 2018), have recently started to talk about **active vs. municipal land policy approach** in a context of land scarcity: meaning that the selection of instruments by public officials has so far often been presented in a functionalist way as if it depends merely on technical choices (Lascoumes & Le Galès, 2005). An active land policy approach, in contrast, involves besides the mere introduction of new rules, in addition, of the strategic activation of the existing rules in force for sustainable settlement transformation. For instance, a city administration might have multiple legal housing instruments in place (e.g. inclusionary zoning, pre-emption rights, housing funding for cooperatives), but still, might fail to implement these measures in a targeted manner (e.g. for the provision of affordable housing for elderly people or marginalized groups). Vice versa, a municipality might have worked for several years on the strategic positioning of its spatial territory (e.g. as climate-neutral city), however, municipal responsitivies lack legal leeway to transform the strategic thoughts to actual measures on the ground (e.g. finances for public land purchase). Therefore, planning scholars have rather recently started to talk about active vs. passive land policy approaches, which are generally distinct as follows:

**Active Municipal Land Policy for Sustainable Housing Development:** refers to a strategy developed, activated, and followed by municipal planning administrations (often in close interaction with other public administrative offices) to effectively reinforce their position in the face of legally powerful landowners in the interest of the public, and in order to address rivalrous land use situations in a targeted manner (Alterman, 1990; Schönig, 2020). Generally spoken, Andreas Hengstermann and Jean-David Gerber (2018) referred to it as "all public decisions and actions aiming to implement politically defined spatial development objectives through changes in the use, distribution and value of land". Hence, active land policy aims to defend a specific democratically predefined policy interest or objective (e.g. affordable housing, climate protection, or adaption) through the deliberate and goal-oriented application of certain instruments or the combination thereof (Gerber et al., 2018). For example, as densification implies a form of planning that goes beyond zoning to deal with complex property rights situations or mosaic-rights of ways, strategic competencies of municipal authorities include e.g. amicable purchase, long-term building leases on public land, public–private contracts, and so on (Hartmann & Spit, 2015; Needham et al., 2018).

Moreover, because property titles give additional power to their holders to shape urban development, public planning authorities too can use them to reinforce their position (Gerber et al., 2017). In the context of urban land scarcity, in addition, municipal planning authorities need to understand the technical challenges of densification. Thus, they must strategically (re)consider neighborhood conflicts, socio-economic population compositions, engaging with stakeholders, initiating projects that align with the long-term vision of the city, *and* they must know the underlying property rights' structure if they aim to densify sustainably, without discrimination, displacement, or social exclusion. Hence, in essence, such an active strategy requires the combination of both—public and private law instruments—and strategic activation of them, in order to not only limit property owners' rights, but also work *with* property rights to enhance the effectiveness of public goal implementation.

In contrast,

**Passive Municipal Land Policy:** refers to a non-goal-oriented strategy followed by municipal planning administrations, which neither involves the targeted introduction of specific instruments, nor a strategic positioning or activation of the rules in force (or both of them lack). Hence, the planning authority does not actively introduce specific instruments or strategic plans to promote development or address specific housing needs. Consequently, the municipal planning authority does often adopt a wait-and-see approach (often also due to missing financial or personnel resources or know-how), relying on external actors (e.g. private developers) to take the initiative in proposing and implementing projects on land within municipal boundaries.

In sum, active municipal land policy differs from a legalistic planning approach in the sense that it involves and brings a more actors-oriented approach to planning (see Chapter 2, theoretical background on "new institutional economics"). It is not only about the mere introduction of instruments, but instead, also about the actors involved, their specific strategies, their personnel or financial resources, and local objectives that decides and influences how, where, and for the benefit of whom spatial (housing) development takes place.

### *3.6.2 The Power of Housing Property Owners as Titleholders*

Property owners (in the housing context often defined as "homeowners") are the target groups of land use policy (Knoepfel & Nahrath, 2007; Knoepfel et al., 2003: 337). As such they play a central role in the institutional regime as they do not only have a right to use the housing stock built on their land, but they also have a contractual obligation to maintenance it. They are entitled to formal property rights and hence have the power to select—through contracting mechanisms—what user-actors have which use rights on the goods and services provided by the housing estate they own (Nicol & Knoepfel, 2008: 166). Moreover, due to their strong position as titleholders, property owners are well-equipped to resist public policy intervention. Consequently, municipal authorities encounter difficulties in implementing their housing and land use schemes due to the powerful resistance of private property owners (Sect. 3.5).

The push for urban density symbolizes enormous economic potential for private owners. They can increase their investment possibilities through the option to (re)construct at central locations and within city boundaries. Either they build on newly created plots or on existing plots where increased density is allowed (Charmes & Keil, 2015; Holman et al., 2015; Touati-Morel, 2015: 606). Urban consolidation presents a real possibility for investors because rent revenue can be enhanced in the short term through targeted redevelopment and upgrading of existing stocks (Brenner, 2009; Brenner & Theodore, 2002; Harvey, 2005; Theodore et al., 2011). Owners can realize an enhanced income stream over an extended time by virtue of their exclusive control over the land and the housing stock that is built on.

Centrality to transport and communications networks, and general proximity within the financial center play a crucial role in this matter, not only because property owners can ask for higher prices for the land and rents because of its accessibility (Theurillat et al., 2014: 1426). Moreover, under scarce land conditions, speculating on future real-estate values becomes a lucrative investment asset (Harvey, 2012: 250–252). The housing stock is being used by institutional investors as a value-enhancing investment opportunity, especially given the current low or even negative interest rate situation in the capital markets. Property owners (particularly institutional investors) consider housing at central locations a safe investment and lucrative speculative object (Aalbers, 2017).

While planners are typically interested in the use value of land and urban housing stocks as they have a greater long-term responsibility to maintain and secure resources use, private investors primarily focus on its exchange value on financial markets (including speculation). Consequently, particularly in liberal states, private owners have a general tendency to consider real-estate assets as an investment, while the municipality appears quite powerless (Gerber et al., 2017: 1686–1699).

In daily practice, three main categories of institutional ownership are differentiated into private foundations, private companies, and private investment funds (for discussion, e.g. Bord, 2006; Csikos, 2008; Theurillat & Crevoisier, 2013; Theurillat et al., 2014). In addition, pension funds and other small private investors are identified as intermediary groups that collect, manage, and invest the deposited funds of their clients (Hübschle et al., 1990). This includes investors such as banks or insurance companies, who invest their money into real estate to deposit their capital. The other three main categories of legal entities belong to the shareholders or the investors, founders, or fund owners who provide the capital for share or foundation creation. Furthermore, there are a variety of small investors (individual homeowners).

Portfolio logic on the financial markets not only guarantees almost instantaneous reallocation of capital as results of the separation between the functions of the economic entrepreneur and the financial investor. Moreover, portfolio value reflects the mimetic behaviors of the shareholders as well as the broader systemic fluctuations in the various financial markets (Theurillat & Crevoisiers, 2013: 2055). In case of private limited companies, general assembly meetings also participate in decision-making. By law, investment funds and private foundations do not have such a "legislative" body, but in case they obtain bank loans, the responsible bank as the creditor is also involved in decision-making processes to ensure returns. All investor types are advised and supported by various private firms of the real-estate industry such as private planning or architecture offices, developers, real-estate managers, construction companies, or rating agencies. The formal owners usually outsource the management and planning of their properties to these specialized firms. Sometimes, these companies are subsidiaries of the landowning firm or belong to the construction companies that build the housing project (Knoepfel et al., 2012).

### *3.6.3 Tenants' Housing Resistance and Decommodification Strategies*

Tenants represent the inhabitants or end-users living in the housing stock (Fig. 3.1; middle row). They either have a right to use a dwelling described in a rent contract or they simply appropriate a housing use that is unregulated (e.g. in the form of squatting or temporary housing) (Nicol & Knoepfel, 2008: 166). In the last two decades, a growing housing crisis has emerged in many cities worldwide, which has strongly influenced how and where residents are able to house (Kemeny, 2006; Lennartz, 2011; Matznetter & Mundt, 2012; Scanlon et al., 2015; Schönig et al., 2017). Many lower-income households no longer had the financial means to become owner-occupiers, to maintain mortgage burdens, or to access public housing (Harloe, 1982: 41).

The transition was further exacerbated by state withdrawal, budget cuts, and a general shift away from subsidized rental housing towards market-based sustenance of housing provision (for discussion, e.g. Aalbers & Holm, 2008; Andersen, 2017; Hackworth & Smith, 2001). This shift included the transformation of the public sector itself, which in many states has become a market-oriented version. For instance, by introducing market principles into public administration (see "new public management" literature, e.g. Dibben & Higgins, 2004; Gerber, 2016; Harvey, 1989; Hughes, 2003; Osborne & Gaebler, 1992). Consequently, the housing supply has generally started to work in the interests of the profit-oriented property sector and not the public. This has led to rising social exclusion and inequality mechanisms in cities (Marcuse, 2009).

However, to counteract trends of social exclusion and gentrification in housing, city residents have initiated resistance strategies that aim towards "decommodification": by definition, decommodification is the counterpart to commodification which can be understood as the action of turning something (housing, goods, people, animals, needs, etc.) into a commodity. For Esping-Andersen (2011), decommodification refers to the degree to which individuals, or families, can uphold a socially acceptable standard of living independent of market participation. In other words: decommodification means the emancipation from market dependencies, and hence is a process that "seeks to get out of the logic of the market, characterized by monetary valuation and exchange, nowhere more prevalent than in property-based economies [and] seeks to leave the 'exchange value' of goods and services to focus more on their 'use value'" (Gerber & Gerber, 2017: 553).

Esping-Andersen considered labor markets, but other scholars included housing, land use, or any other markets in their decommodification research (Kadi & Ronald, 2014; Kolocek, 2017). Holm (2006), for instance, added that such emancipation from housing market dynamics occurs in three different areas: money (through funding and subsidies), property (when building permits are equipped with certain duties for the developer), and rights. Examples of decommodification through rights are legal protection from fast-rising rents or eviction, both typical tenancy law instruments.

In housing, moreover, decommodification stands for the residents' attempt to create strong social entitlements and for the citizens' degree of immunization from market dependencies (Kadi & Ronald, 2014: 270). The central goal is "to provide every person with housing that is affordable, adequate in size and of decent quality, secure in tenure, and located in a supportive neighborhood of choice, with recognition of the special housing problems confronting oppressed groups" (Achtenberg & Marcuse, 1986: 476).

In reality and on the daily ground, residents often address such decommodification attempts through NIMBY (Not-in-my-backyard) opposition. They experience asymmetric distribution of power and a loss of social stability resulting in social exclusion, polarization, and gentrification (for discussion, e.g. Burbank et al., 2000; Dear, 1992; Kübler, 1995; Pendall, 1999; Scally & Tighe, 2015; Searle & Filion, 2011). This NIMBYism can take different forms, for example, street rallies, petitions, social movements, or neighborhood-wide objections against city (re)development, (re)vitalization, or upgrading projects (Holm & Kuhn, 2010; Peck & Tickell, 2002). But also practices of cooperation, mutual aid, solidarity, as well as horizontality, non-hierarchy, and equality. In critical urban geography literature, such resistance strategies are generally summarized under the umbrella term "right-to-the-city" movements (for discussion, e.g. Castells, 1977; Fainstein & Fainstein, 1985; Harvey, 1973, 2008; Lefebvre, 1970; Lowe, 1986; Mayer, 2003; Zukin, 1982). Residents participating in such movements particularly seek affordable housing provision, a decent standard of living, and/or protection against displacement or income inequality. Moreover, many of them aim to raise awareness that housing should not be considered a commodity but a source of basic need satisfaction upon which people depend absolutely (Hackworth & Smith, 2001; Scally & Tighe, 2015).

For example, many tenant activist groups have started to find cooperative housing associations that bought dwellings (or land) to build new homes. As collective owners, they are directly involved in the collective management of their homes with the freedom to physically modify their individual dwellings as they wish. Rents are set at a level necessary to service any debts incurred (cost rent principle) and build up an equity share in the property so that when a tenant leaves, they receive capital returns based on their share (Hodkinson, 2012). By removing the land and the housing stock from the private property market and controlling its use in perpetuity and mutuality, collective ownership becomes as attractive as individual home ownership. Any speculative and inflationary forces driving up the rents for the existing community can be stopped while any increase in value stays with the local community (Ward, 1974: 131). This overview of the literature indicates *how* tenants can develop effective resistance strategies to defend their own socially sustainable housing interests in dense city environments.

## 3.7 Hypotheses: Governance Mechanisms Leading to Socially Sustainable Housing Outcomes in Densifying Cities

The IRR analytical approach makes it possible to analyze in detail how institutional rules are used by different resource users and how they contribute to potential positive outcomes in terms of social sustainability in housing. Densification projects "that are ecologically viably, but socially not accepted as places in which to live, work or interact cannot be acknowledged sustainable" (Bramley et al., 2009: 2125). Given this potential for trade-offs among goals, a choice must be made as to which objectives should receive priority and hence greater weight in the densification process (Barbier, 1987: 104). However, these trade-offs are identified as the results of power games among actors. Densification objectives never get implemented on a one-to-one basis. Rather, power games influence the implementation process (e.g. Fainstein & Fainstein, 1979; Flyvbjerg, 1998; Friedmann, 1998).

By combining approaches from public policy analysis (planning as a public policy), and new institutional economics (property rights) with political ecology (power), this book enables to capture these power games among actors. Powerful actors are those who know how to influence the strategies and goals of other actors in a targeted manner to promote and protect their own values, needs, and objectives. By making explicit the governance mechanisms of possible sustainability trade-offs and power games in densification procedures, this research project contributes to neoinstitutionalist political ecology research. It recognizes that many different resource users can come into conflict with each other and allows for a systematic analysis of how various actors behave in response to a specific socio-political setting.

In the following paragraph, I formulate five broad working hypotheses on the local governance mechanisms contributing to socially sustainable housing in dense city areas (Fig. 3.2). I explain the power relations between actors involved in consolidation projects to understand the reasons for potential trade-offs between economic, environmental, *and* social dimensions. By nature, working hypotheses present a simplified statement about a complex reality. They are not to be "tested" in a quantitative manner (Gläser & Laudel, 2010); rather they are statements about possible causalities that guide and structure the research process.

**SQ1:** How do institutional rules affect the outcomes of densification in terms of social sustainability in housing?

**H1: Institutional Incoherence Between Landowners' and Tenants' Rights Prevents Social Sustainability in Housing**

*In the context of densification, the (Swiss) legal regime in force is not able to secure the provision of all goods and services provided by the housing stock (e.g. energy, capital investment, affordable living space). The policy shift towards densification has changed the balance of power between different users, particularly between landowners and tenants: while the business interests of owner-actors remain strongly protected by law (through property rights), tenants are not in the legal position to be heard even though the land use conditions have changed. This institutional incoherence between*

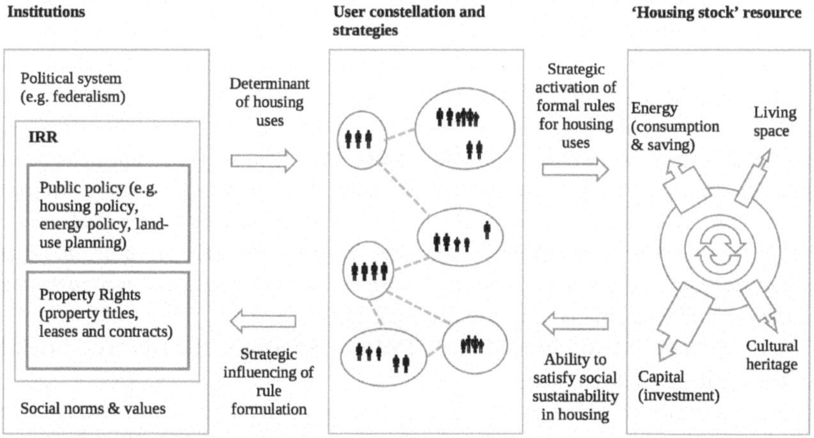


**Fig. 3.2** The Institutional Resource Regime postulates a causal relationship (hypotheses 1–5) between three main variables: the condition of the housing-stock resource (dependent variable), actors (users) (intermediary variable), and institutions (independent variable) (Knoepfel et al., 2007; Gerber & Debrunner, 2022; Gerber et al., 2009, 2020: 157)

*landowners' and tenants' rights leads to the neglect of tenants' affordable housing needs while owners' profitability objectives take the upper hand.*

**SQ2:** What use strategies do actors (owners and non-owners) follow to contribute to socially sustainable housing in a densifying city?

**H2: Resistance Power of Landowners Prevents Social Sustainability in Housing**

*In the context of densification, property owners (and the private real-estate industries working on their behalf) strategically activate their rights to private property to increase profit making. They use their rights to set the rents according to market prices and to increase the land rent. Furthermore, through a strategic coalition of political forces (in legislative and executive committees), they constantly prevent the formulation or introduction of more restrictive regulations that could potentially harm their profit interests in housing (re)development. Thereby, they divert the original intensions of densification policy objectives (green growth) by using the legal provisions to achieve goals other than those originally planned by the law (Eco-Business).*

**H3: Effective Tenants' Grassroots Resistance Leads to Social Sustainability in Housing**

*Densification triggers residents' grassroots resistance to social sustainability goals in housing as well as political and public debates on alternative densification models because tenants suffer the negative consequences of the process (rising rents, social exclusion). Tenants strategically participate to NIMBY-opposition (e.g. public rallies, legal objections, formal petitions) to increase the socio-political pressure on governments and owner-actors to obtain measures against dismissal and displacement (e.g. compensation payments, secure political support, reduction of local costs). If those entities do not agree to compromise with the tenants' claims, landowners and public authorities riskthat densification as public policy objective as a whole will fail or come to a standstill because of tenants' social resistance against and missing social acceptance for (re)development.*

**H4: Active Municipal Land Policy Strategy Leads to Social Sustainability in Housing**

*Densification forces municipal authorities to activate regulations and policy instruments that promote social sustainability goals in housing that are otherwise being put aside. Public-administrative actors are in the key position to strategically intervene in private development interests because land use planning alone is unable to control how owners use their land in the interest of the public. They can prevent tenants' social exclusion, but they must therefore know how to densify. To do so, they need to act strategically and activate public and private law instruments, which do not always need to limit property owners' rights (e.g. through zoning), but also work with property rights (e.g. through public land acquisition). All the finesse and strategic competencies of planning administrations is required to implement planning measures that promote social inclusion because landowners have the power to defend the status quo due to strong veto rights.*

**SQ3:** How does the implementation of densification objectives impact socially sustainable housing outcomes?

**H5: The Business of Densification**

*Densification—as a core objective of (Swiss) spatial planning policy—leads to the neglect of the social pillar of sustainability in housing because owner-actors have realized that the shift towards densification comes with real business opportunities rather than profit restrictions. Today, they acknowledge that densification is a profitable investment opportunity since more rent revenue can be realized at central locations through redevelopment and upgrading of existing stocks. Simultaneously, public authorities have started to promote the "Business of Densification" too since they have noticed that property owners—due to their strongly protected property title—are in a strong legal position to resist socially sustainable densification efforts. However, within this trade-off of densification objectives, the three sustainability dimensions fallout of balance. Powerful economic actors (including landowners, real-estate industries) and public authorities put social criteria (e.g. social mixing, tenure security) on the back burner, while economic and ecological criteria of densification become more prioritized.*

## References


Aalbers, M. B., & Holm, A. (2008). Privatising social housing in Europe: The cases of Amsterdam and Berlin. In K. Adelhof, B. Glock, J. Lossau, & M. Schulz (Eds.), *Urban trends in Berlin and Amsterdam* (pp. 12–23). Geographisches Institut der Humboldt-Universität.

Aalbers, M. B. (2017). The variegated financialization of housing. *International Journal of Urban and Regional Research, 41*, 542–554.

Achtenberg, E. P., & Marcuse, P. (1986). Toward the decommodification of housing. In R. Bratt, C. Hartman, & A. Meyerson (Eds.), *Critical perspectives on housing* (pp. 474–483). Temple University Press.

Alexander, D., & Tomalty, R. (2002). Smart growth and sustainable development: Challenges, solutions and policy directions. *Local Environment, 7*(4), 397–409.

Alexander, E. R., & Faludi, A. (1989). Planning and plan implementation: Notes on evaluation criteria. *Environment & Planning b: Planning & Design, 16*(2), 127–140.

Alterman, R. (1990). *Private supply of public services: Evaluation of real state exactions, linkage, and alternative land policies*. University Press.

Alterman, R. (2007). More than land assembly: Land readjustment for the supply of urban public services. In Y. Hong & B. Needham (Eds.), *Analyzing land readjustment* (pp. 57–88). Lincoln Institute of Land Policy.

Andersen, H. T. (2017). Copenhagen: The social costs of urban renewal. In R. Cucca & R. Costanzo (Eds.), *Unequal cities* (pp. 138–154). Routledge.

Baer, W. C. (1997). General plan evaluation criteria: An approach to making better plans. *Journal of the American Planning Association, 63*, 329–344.

Balmer, I., & Bernet, T. (2015). Housing as a common resource? Decommodification and self-organization in housing—Examples from Germany and Switzerland. In M. Dellenbaugh, M. Kip, M. Bienick, A. K. Müller, & M. Schwegmann (Eds.), *Urban commons: Moving beyond state and market* (154th Ed., pp. 178–195). Birkhäuser.

Balmer, I., & Gerber, J.-D. (2017). 'Why are housing cooperatives successful? Insights from Swiss affordable housing policy. *Housing Studies, 33*, 361–385.

Baptista, I., & O'Sullivan, E. (2008). The role of the state in developing homeless strategies: Portugal and Ireland in comparative perspective. *European Journal of Homelessness, 2*, 25–44.

Barbier, E. B. (1987). The concept of sustainable economic development. *Environmental Conservation, 14*(2), 101–110.

Barry, B. (1991). Sustainability and intergenerational justice. In A. Dobson (Ed.), *Fairness and futurity* (pp. 230–250). Oxford University Press.

Basiago, A. D. (1999). Economic, social, and environmental sustainability in development theory and urban planning practice. *The Environmentalist, 19*(2), 145–161.

Bathelt, H., & Glückler, J. (2005). Resources in economic geography: From substantive concepts towards a relational perspective. *Environment and Planning A, 37*(9), 1545–1563.

Beer, A., Bentley, R., Baker, E., Mason, K., Mallett, S., Kavanagh, A., & LaMontagne, T. (2016). Neoliberalism, economic restructuring and policy change: Precarious housing and precarious employment in Australia. *Urban Studies, 53*(8), 1542–1558. https://doi.org/10.1177/0042098015596922

Bengtsson, B. (2001). Housing as a social right: Implications for welfare state theory. *Scandinavian Political Studies, 24*(4), 255–275.

Benjaminsen, L., & Dyb, E. (2008). The effectiveness of homeless policies—Variations among the Scandinavian countries. *European Journal of Homelessness, 2*, 45–68.

Bhatti, M. (2001). Housing/futures? The challenge from environmentalism. *Housing Studies, 16*(1), 39–52.
Bhatti, M., & Dixon, A. (2003). Special focus: Housing, environment and sustainability. *Housing Studies, 18*(4), 501–504.
Biart, M. (2002). Social sustainability as part of the social agenda of the European Community. In T. Ritt (Ed.), *Soziale Nachhaltigkeit: Von der Umweltpolitik zur Nachhaltigkeit?* (pp. 5–10). Arbeiterkammer Wien, Informationen zur Umweltpolitik 149.
Bibby, P., Henneberry, J., & Halleux, J.-M. (2018). Under the radar? "Soft" residential densification in England, 2001–2011. *Environment and Planning B: Urban Analytics and City Science, 0*(0), 1–17.
Bord, M. (2006) *Analyse juridique des forms de propriéte et de propriétaires collectifs en Suisse*. Lausanne.
Bramley, G., Dempsey, N., Power, S., Brown, C., & Watkins, D. (2009). Social sustainability and urban form: Evidence from five British cities. *Environment and Planning A, 41*(9), 2125–2142.
Bramley, G., & Power, S. (2009). Urban form and social sustainability: The role of density and housing type. *Environment and Planning b: Planning and Design, 36*(1), 30–48.
Breheny, M. (1992). *Sustainable development and urban form*. Pion.
Brenner, N., & Theodore, N. (2002). *Spaces of neoliberalism*. Blackwell.
Brenner, N. (2009). Cities and territorial competitiveness. In C. Rumford (Ed.), *Handbook of European studies* (pp. 442–463). Sage.
Brenner, N., Marcuse, P., & Mayer, M. (2012). *Cities for people, not for profit: Critical urban theory and the right to the city* (edited by N. Brenner, P. Marcuse, & M. Mayer). Routledge.
Broitman, D., & Koomen, E. (2015). Residential density change: Densification and urban expansion. *Computers, Environment and Urban Systems, 54*, 32–46.
Bromley, R. D. F., Tallon, A. R., & Thomas, C. J. (2005). City centre regeneration through residential development: Contributing to sustainability. *Urban Studies, 42*(13), 2407–2429.
Budd, W., Lovrich, N., Pierce, J. C., & Chamberlain, B. (2008). Cultural sources of variations in US urban sustainability attributes. *Cities, 25*(5), 257–267.
Burbank, M. J., Heying, C. H., & Andranovich, G. (2000). Antigrowth politics or piecemeal resistance? Citizen opposition to olympic-related economic growth. *Urban Affairs Review, 35*(3), 334–357.
Burton, E. (2003). Housing for an urban renaissance: Implications for social equity. *Housing Studies, 18*(4), 537–562.
Campbell, S. (1996). Green cities, growing cities, just cities?: Urban planning and the contradictions of sustainable development. *Journal of the American Planning Association, 62*(3), 296–312.

Cardoso, R., Sobhani, A., & Meijers, E. (2021). The cities we need: Towards an urbanism guided by human needs satisfaction. *Urban Studies, 59*(13), 2638–2659. https://doi.org/10.1177/00420980211045571
Castells, M. (1977). *The urban question*. MIT Press.
Cernea, M. M. (1993). Culture and organization—The social sustainability of induced development. *Sustainable Development, 1*(2), 18–29.
Charmes, E., & Keil, R. (2015). The politics of post-suburban densification in Canada and France. *International Journal of Urban and Regional Research, 39*, 581–602.
Chiu, R. L. H. (2003). Social sustainability, sustainable development and housing development: The experience of Hong Kong. In R. Forrest & J. Lee (Eds.). *Housing and social change. East-West perspectives*. Routledge.
Chiu, R. L. H. (2004). Socio-cultural sustainability of housing: A conceptual exploration. *Housing, Theory and Society, 21*(2), 65–76.
Christen, M., & Schmidt, S. (2012). A formal framework for conceptions of sustainability—A theoretical contribution to the discourse in sustainable development. *Sustainable Development, 20*(6), 400–410.
Colantonio, A., & Dixon, T. (2009). *Measuring socially sustainable urban regeneration in Europe* (pp. 1–129). Oxford Institute for Sustainable Development (OISD). https://doi.org/10.1002/9780470690604
Cook, B., & Kothari, U. (2001). *Participation: The new tyranny?* Zed Books.
Crabtree, L. (2006). Sustainability begins at home? An ecological exploration of sub/urban Australian community-focused housing initiatives. *Geoforum, 37*(4), 519–535.
Csikos, P. (2008). *Les stratégies foncières des grands propriétaires fonciers coll les cas des sociétés immobilières, fonds de placement immobilier et fondation placement*. Chavannes-près-Renens.
Curdes, G. (1995). *Stadtstrukturelles Entwerfen*. Kohlhammer.
Dave, S. (2010). High urban densities in developing countries: A sustainable solution? *Built Environment, 36*(1), 9–27.
Davidson, M. (2010). Social sustainability and the city. *Geography Compass, 4*(7), 872–880.
Dear, M. (1992). Understanding and overcoming the NIMBY syndrome. *Journal of the America Planning Association, 58*(3), 288–300.
Dempsey, N., Bramley, G., Power, S., & Brown, C. (2009). The social dimension of sustainable development: Defining urban social sustainability. *Sustainable Development, 19*, 289–300.
Dibben, P., & Higgins, P. (2004). New public management: Marketisation, managerialism and consumerism. In P. Dibben, G. Wood, & I. Roper (Eds.), *Contesting public sector reforms. Critical perspectives, international debates*. Palgrave-Macmillan.

Dolsak, N., & Ostrom, E. (2003). *The commons in the new millennium*. The MIT Press.

Elkin, T., McLaren, D., & Hillman, M. (1991). *Reviving the city: Towards sustainable urban development*. Friends of the Earth.

Elkington, J. (1999). *Cannibals with forks*. Capstone Publishing.

Elliot, J. A. (1999). *Introduction to sustainable develoment* (2nd ed.). Taylor & Francis.

Esping-Andersen, G. (2011). Welfare regimes and social stratification. *Journal of European Social Policy, 25*(1), 124–134. https://doi.org/10.1177/0958928714556976

Evans, B., Joas, M., Sundback, S., & Theobald, K. (2006). Governing local sustainability. *Journal of Environmental Planning and Management, 49*(6), 849–867.

Evans, J. P., & Jones, P. (2008). Rethinking sustainable urban regeneration: Ambiguity, creativity, and the shared territory. *Environment and Planning A, 40*(6), 1416–1434.

Fainstein, S. F. (2010). *The just city*. Cornell University Press.

Fainstein, S. S., & Fainstein, N. I. (1985). Economic restructuring and the rise of urban social movements. *Urban Affairs Quarterly, 21*(2), 187–206.

Fainstein, S. S. (2001). Competitiveness, cohesion, and governance: Their implications for social justice. *International Journal of Urban and Regional Research, 25*(4), 884–888.

Fainstein, N. I., & Fainstein, S. S. (1979). New debates in urban planning. *International Journal of Urban and Regional Research, 3*(1–4), 381–403.

Flyvbjerg, B. (1998). *Rationality and power democracy in practice*. University of Chicago Press.

Foladori, G. (2005). Advances and limits of social sustainability as an evolving concept. *Canadian Journal of Development Studies, 26*(3), 501–510.

Friedmann, J. (1998). Planning theory revisited. *European Planning Studies, 6*(3), 245–253.

Fürst, D., & Scholles, F. (2001). *Handbuch Theorien und Methoden der Raum- und Umweltplanung* (2nd ed.). Dorothea Rohn.

Gerber, J.-D., Koepfel, P., Nahrath, S., & Varone, F. (2009). Institutional resource regimes: Towards sustainability through the combination of property-rights theory and policy analysis. *Ecological Economics, 68*(3), 798–809.

Gerber, J.-D. (2016). The managerial turn and municipal land-use planning in Switzerland—Evidence from practice. *Planning Theory and Practice, 17*(2), 192–209.

Gerber, J.-D., & Gerber, J. F. (2017). Decommodification as a foundation for ecological economics. *Ecological Economics, 131*, 551–556.

Gerber, J.-D., Nahrath, S., & Hartmann, T. (2017). The strategic use of time-limited property rights in land-use planning: Evidence from Switzerland. *Environment and Planning A, 49*(7), 1684–1703.

Gerber, J.-D., Hartmann, T., & Hengstermann, A. (2018). *Instruments of land policy—Dealing with scarcity of land* (edited by J.-D. Gerber, T. Hartmann, & A. Hengstermann). Routledge.

Gerber, J.-D. (2018). A Swiss perspective on time-limited property rights: Strategic use for active land policy. In J.-D. Gerber, T. Hartmann, & A. Hengstermann (Eds.), *Instruments of land policy—Dealing with scarcity of land* (pp. 260–264). Routledge.

Gerber, J.-D., Lieberherr, E., & Knoepfel, P. (2020). Governing contemporary commons: The institutional resource regime in dialogue with other policy frameworks. *Environmental Science and Policy, 112*(6), 155–163.

Gerber, J.-D., & Debrunner, G. (2022). Power and planning. A theory-informed discussion of a large housing development project in Zurich, Switzerland. *Land Use Policy*. https://doi.org/10.1016/j.landusepol.2022.106400

Gläser, J., & Laudel, G. (2010). *Experteninterviews und qualitative inhaltsanalyse* (4th ed.). VS Verlag für Sozialwissenschaften.

Goulet, D. (1971). *The cruel choice: A new concept in the theory of development*. Atheneum.

Graham, J., Amos, B., & Plumptre, T. (2003). *Governance principles for protected areas in the 21st century*. Institute on Governance.

Griessler, E., & Littig, B. (2005). Social sustainability: A catchword between political pragmatism and social theory Beate Littig * and Erich Grießler. *International Journal Sustainable Development, 8*(1/2), 65–79.

Hackworth, J., & Smith, N. (2001). The changing state of gentrification. *Royal Dutch Geographical Society KNAG, 6*(9), 464–477.

Haller, T., Acciaioli, G., & Rist, S. (2016). Constitutionality: Conditions for crafting local ownership of institution-building processes. *Society and Natural Resources, 29*(1), 68–87.

Hargreaves, A. (2004). Building communities of place: Habitual movement around significant places. *Journal of Housing and the Built Environment, 19*(1), 49–65.

Harloe, M. (1982). Towards the decommodification of housing?: A comment on council house sales. *Critical Social Policy, 2*(4), 39–42.

Hartmann, T., & Spit, T. (2015). Dilemmas of involvement in land management—Comparing an active (Dutch) and a passive (German) approach. *Land Use Policy, 42*, 729–737.

Harvey, D. (1973). *Social justice and the city*. Hodder and Stoughton.

Harvey, D. (1989). From managerialism to entrepreneurialism: The transformation in urban governance in late capitalism and geography. *GeografiskaAnnaler, 71*(1), 3–17.

Harvey, D. (2005). *A brief history of neoliberalism*. Oxford University Press.
Harvey, D. (2008). The right to the city. *New Left Review, 53*, 23–40.
Harvey, D. (2012). *Rebel cities: From the right to the city to the urban revolution*. Verso.
Heidenreich, F. (2011). *Theorien der Gerechtigkeit*. UTB GmbH.
Hengstermann, A., & Gerber, J.-D. (2015). Towards active land policy—The new Swiss Spatial Planning Act. *fub, 6*, 241–250.
Hengstermann, A. (2019). *Von passiver Bodennutzungsplanung zu aktiver Bodenpolitik—Eine Fallstudie zur Wirksamkeit von bodenpolitischen Instrumenten anhand von Lebensmittel-Discountern* (1st ed.). Springer VS.
Hodkinson, S. (2012). The return of the housing question. *Ephemera: Theory & Politics in Organization, 12*(4), 423–444.
Holm, A. (2006). Urban renewal and the end of social housing: The roll out of neoliberalism in east Berlin's prenzlauer berg. *Social Justice, 33*(3), 114–128.
Holm, A., & Kuhn, A. (2010). Squatting and urban renewal: The interaction of squatter movements and strategies of urban restructuring in Berlin. *International Journal of Urban and Regional Research*, 1–15.
Holman, N., Mace, A., Paccoud, A., & Sundaresan J. (2015). Coordinating density: Working through conviction, suspicion and pragmatism. *Progress in Planning, 101*, 1–38. Elsevier Ltd.
Hübschle, J., Schultz, H. R., Hager, A., & Müller, U. (1990). *Sozialversicherungssparen und Immobilienmarkt—Bericht 64 des Nationalen Forschungsprogrammes "Boden."* IRB Verlag.
Hughes, O. E. (2003). *Public management and administration: An introduction*. Palgrave Macmillan.
IFHP [International Federation for Housing and Planning]. (2019). *The Social cities index: A tool for urban leaders*.
Jabareen, Y. (2008). A new conceptual framework for sustainable development. *Environment, Development and Sustainability, 10*(2), 179–192.
Jacobs, M. (1992). *The green economy: Environment, sustainable development and the politics of the future* (edited by M. Jacobs). Pluto Press.
Jenks, M., Burton, E., & Williams, K. (1996). *The compact city: A sustainable urban form?* (1st ed.). Routledge.
Jonkman, A. (2019). *Distributive justice of housing in Amsterdam*. University of Amsterdam.
Kadi, J., & Ronald, R. (2014). Market-based housing reforms and the "right to the city": The variegated experiences of New York, Amsterdam and Tokyo. *International Journal of Housing Policy, 14*(3), 268–292.
Karlsson, J. (2018). The sustainable city—Promoting socially sustainable cities through civil dialogue. *Provocatio, 3*, 7–18.
Kébir, L. (2010). Pour une approche institutionnelle et territoriale des ressources. Bruxelles: Peter Lang. In M. Maillefert, O. Petit, & S. Rousseau (Eds.),

*Ressources, patrimoine, territoires et développement durable* (pp. 69–86). P.I.E. Peter Lang.

Kemeny, J. (2001). Comparative housing and welfare: Theorising the relationship. *Journal of Housing and the Built Environment, 16*(1), 53–70. https://doi.org/10.1023/A:1011526416064

Kemeny, J. (2006). Corporatism and housing regimes. *Housing, Theory and Society, 23*(1), 1–18.

Ketschau, T. J. (2015). Social justice as a link between sustainability and educational sciences. *Sustainability, 7*(11), 15754–15771.

Khan, M. A. (1995). Sustainable development: The key concepts, issues and implications. *Sustainable Development, 3*, 63–69.

Knoepfel, P., Kissling-Näf, I., & Varone, F. (2001). *Institutionelle Regime für natürliche Ressourcen: Boden, Wasser und Wald im Vergleich*. Helbing & Lichtenhahn.

Knoepfel, P., Kissling-Näf, I., & Varone, F. (2003). *Institutionelle Regime natürlicher Ressourcen in Aktion*. Helbing & Lichtenhahn.

Knoepfel, P., & Nahrath, S. (2007). Environmental and spatial development policy. In U. Klöti, P. Knoepfel, & H. Kriesi (Eds.), *Handbook of Swiss politics* (pp. 705–733). NZZ Verlag.

Knoepfel, P., Nahrath, S., & Varone, F. (2007). Institutional regimes for natural resources: An innovative theoretical framework for sustainability. In P. Knoepfel (Ed.), *Environmental policy analyses. Learning from the past for the future. 25 years of research* (pp. 455–506). Springer.

Knoepfel, P., Larrue, C., Varone, F., & Hill, M. (2011). *Public policy analysis*. Policy Press.

Knoepfel, P., Csikos, P., Gerber, J.-D., & Narath, S. (2012). Transformation der Rolle des Staates und der Grundeigentümer in städtischen Raumentwicklungsprozessen im Lichte der nachhaltigen Entwicklung. *Nomos Verlagsgesellschaft, 53*(3), 414–443.

Kolocek, M. (2017). *The human right to housing in the face of land policy and social citizenship: A global discourse analysis*. Palgrave Macmillan.

Krueger, R., & Agyeman, J. (2005). Sustainability schizophrenia or "actually existing sustainabilities?" Toward a broader understanding of the politics and promise of local sustainability in the US. *Geoforum, 36*(4), 410–417.

Krueger, R., & Gibbs, D. (2007). *The sustainable development paradox: Urban political economy in the United States and Europe*. Guilford.

Kübler, D. (1995). Problèmes de mise en œuvre de la politique sociale en milieu urbain: l'exemple des services médico-sociaux pour consommateurs de drogues. *Swiss Political Science Review, 1*(4), 99–120.

Kyttä, M., Broberg, A., Tzoulas, T., & Snabb, K. (2013). Towards contextually sensitive urban densification: Location-based softGIS knowledge revealing perceived residential environmental quality. *Landscape and Urban Planning, 113*, 30–46.

Langhelle, O. (2000). Sustainable development and social justice: Expanding the Rawlsian framework of global justice. *Environmental Values, 9*(3), 295–323.

Lascoumes, P., & Le Galès, P. (2005). Gouverner par les instruments. *Presses de Sciences Po (P.F.N.S.P.).*

Laurian, L., Crawford, J., & Day, M. (2010). Evaluating the outcomes of plans: Theory, practice, and methodology. *Environment and Planning B: Planning and Design, 37*, 740–757.

Lee, B. A., & Evans, M. (2020). 'Forced to move: Patterns and predictors of residential displacement during an era of housing insecurity. *Social Science Research*. https://doi.org/10.1016/j.ssresearch.2020.102415

Lefebvre, H. (1970). *Le droit à la ville*. Antropos.

Lennartz, C. (2011). Power structures and privatization across integrated rental markets: Exploring the cleavage between typologies of welfare regimes and housing systems. *Housing, Theory & Society, 28*(4), 342–359.

Lombard, M. (2021). The experience of precarity: Low-paid economic migrants' housing in Manchester. *Housing Studies*, 1–19. https://doi.org/10.1080/02673037.2021.1882663

Lowe, S. (1986). *Urban social movements. The city after Castells*. St. Martin's Press.

Manzi, T. (2010). *Social sustainability in urban areas: Communities, connectivity and the urban fabric*. Earthscan.

Marcuse, P. (1998). Sustainability is not enough. *Environment and Urbanization, 10*(2), 103–111.

Marcuse, P. (2009). A critical approach to the subprime mortgage crises in the United States: Rethinking the public sector in housing. *City and Community, 8*(3), 351–356.

Marshall, T. H. (1964). Citizenship and social class. In Marshall, T. H. (Ed.), *Class citizenship and social development: Essays by T. Marshall*. Green Press.

Matznetter, W., & Mundt, A. (2012). Housing and welfare regimes. In D. F. Clapham, W. A. V. Clark, & K. Gibb (Eds.), *Housing studies* (1st ed., pp. 274–294). SAGE.

Mayer, M. (2003). The onward sweep of social capital: Causes and consequences for understanding cities, communities and urban movements. *International Journal of Urban and Regional Research, 27*(1), 110–132.

McCrea, R. O. D., & Walters, P. (2012). Impacts of urban consolidation on urban liveability: Comparing an inner and outer suburb in Brisbane, Australia. *Housing, Theory and Society, 29*(2), 190–206.

Miller, D. (1978). Democracy and social justice. *British Journal of Political Science, 8*(1), 1–19.

Mitlin, D., & Satterthwaite, D. (1996). Sustainable development and cities. In C. Pugh (Ed.), *Sustainability, the environment and urbanisation* (pp. 23–62). Earthscan Publications Limited.

Mulliner, E., Smallbone, K., & Maliene, V. (2013). An assessment of sustainable housing affordability using a multiple criteria decision making method. *Omega, 41*, 270–279.

Murphy, K. (2012). The social pillar of sustainable development: A literature review and framework for policy analysis. *Sustainability: Science Practice, & Policy, 8*(1), 15–29.

Næss, P., & Saglie, I. (2019). Ecological modernisation: Achievements and limitations of densification. In S. Davoudi, R. Cowell, I. White, & H. Blanco (Eds.), *The Routledge companion to environmental planning* (pp. 63–72). Routledge.

Needham, B., & Verhage, R. (1998). The effects of land policy: Quantity as well as quality is important. *Urban Studies, 35*(1), 25–44.

Needham, B. (2006). *Planning, law and economics*. Routledge.

Needham, B. (2014). *Dutch land-use planning: The Principles and the Practice.* Ashgate.

Needham, B., Buitelaar, E., & Hartmann, T. (2018). *Planning law and economics—The rules we make for using land* (2nd ed.). Routledge.

Newman, K., & Wyly, E. (2006). The right to stay put, revisited: Gentrification and resistance to displacement in New York City. *Urban Studies, 43*(1), 23–57.

Nicol, L. A., & Knoepfel, P. (2008). Institutional regimes for sustainable collective housing stocks. *Swiss Political Science Review, 14*(1), 157–180.

Nicol, L. A. (2011). The role of institutional regimes in motivating change for sustainable housing. *Building Research and Information, 39*(5), 459–472.

Nicol, L. A. (2013). *Sustainable collective housing. Policy and practice for multi-family dwellings*. Routledge.

North, D. C. (1994). Economic performance through time. *The American Economic Review, 84*(3), 359–368.

Norton, R. K. (2005). More and beter local planning: State-mandated local planning in Coastal North Carlolina. *Journal of the American Planning Association, 71*, 55–71.

Oliveira, V., & Pinho, P. (2011). Bridging the gap between planning evaluation and programme evaluation: The contribution of the PPR methodology. *Evaluation, 17*(3), 293–307.

Osborne, D., & Gaebler, T. (1992). *Reinventing government: How the entrepreneurial spirit is transforming the public sector*. Plume.

O'Sullivan, E., & de Decker, P. (2007). Regulating the private rental housing market in Europe. *European Journal of Homelessness, 1*, 95–117.

Oxford Institute for Sustainable Development. (2007). Social sustainability: An exploratory analysis of its definition, assessment methods, metrics and tools. In *Measuring Social Sustainability: Best Practice from Urban Renewal in the EU*.

Peck, J., & Tickell, A. (2002). Neoliberalizing space. *Antipode*, 380–404.

Pendall, R. (1999). Opposition to housing: NIMBY and beyond. *Urban Affairs Review, 35*(1), 112–136.

Polanyi, K. (1957). *The great transformation*. Beacon Press.

Polese, M., & Stren, R. (2000). *The social sustainability of cities. Diversity and the management of change* (edited by M. Polese & R. Stren). University of Toronto Press.

Portney, K. (1994). Environmental justice and sustainability: Is there a critical nexus in the case of waste disposal or treatment facility siting? *Fordham Urban Law Journal, 11*, 827–839.

Priemus, H. (2005). How to make housing sustainable? The Dutch experience. *Environment and Planning B: Planning and Design, 32*(1), 5–19.

Rawls, J. (1972). *A theory of justice*. Harvard University Press.

Redclift, M. R. (2005). An oxymoron comes of age. *Sustainable Development, 13*(4), 212–227.

Rérat, P. (2012). The new demographic growth of cities: The case of reurbanisation in Switzerland. *Urban Studies, 49*, 1107–1125.

Robbins, P. (2004). *Political ecology: A critical introduction*. Blackwell Publishing.

Ronald, R. (2013). Housing and welfare in Western Europe: Transformations and chal-lenges for the social rented sector. *LHI Journal of Land, Housing, and Urban Affairs, 4*(1), 1–13. https://doi.org/10.5804/lhij.2013.4.1.001

Rousseau, J.-J. (1762). *Du contrat social ou principes du droit politique*. H. Denhardt.

Rudolf, S. C., Kienast, F., & Hersperger, A. M. (2018). Planning for compact urban forms: Local growth-management approaches and their evolution over time. *Journal of Environmental Planning and Management, 61*(3), 474–492.

Sachs, I. (1999). Social sustainability and whole development: Exploring the dimensions of sustainable development. In B. Egon & J. Thomas (Eds.), *Sustainability and the social sciences: A cross-disciplinary approach to integrating environmental considerations into theoretical reorientation*. Zed-Books.

Savini, F. (2011). The endowment of community participation: Institutional settings in two urban regeneration projects. *International Journal of Urban and Regional Research, 35*(5), 949–968.

Scally, C. P., & Tighe, J. R. (2015). Democracy in action?: NIMBY as impediment to equitable affordable housing siting. *Housing Studies, 30*(5), 749–769.

Scanlon, K., Fernandez Arrigoitia, M., & Whitehead, C. (2015). Social housing in Europe. *European Policy Analysis, 17*(7), 1–12.

Schmidt-Thomé, K., Haybatollahi, M., Kyttä, M., & Korpi, J. (2013). The prospects for urban densification: A place-based study. *Environmental Research Letters, 8*(2), 1–11.

Schönig, B., Rink, D., Gardemin, D., & Holm, A. (2017). Paradigmenwechsel in der kommunalen Wohnungspolitik? Variationen kommunalisierter Wohnungspolitik im transformierten Wohlfahrtsstaat. In M. Barbehön & S. Münch (Eds.), *Variationen des Städtischen—Variationen lokaler Politik* (pp. 25–62). Springer VS.

Schönig, B. (2020). Paradigm shifts in social housing after welfare-state transformation: Learning from the German experience. *International Journal of Urban and Regional Research*, 1–18. https://doi.org/10.1111/1468-2427.12914

Searle, G., & Filion, P. (2011). Planning context and urban intensification outcomes: Sydney versus Toronto. *Urban Studies, 48*(7), 1419–1438.

Sen, A. (1999). *Development freedom, development freedom* (edited by Amartya Sen). Oxford University Press.

Shahab, S., Clinch, J. P., & O'Neill, E. (2019). Impact-based planning evaluation: Advancing normative criteria for policy analysis. *Environment and Planning b: Urban Analytics and City Science, 46*(3), 534–550.

Slater, T. (2015). Planetary rent gaps. *Antipode, 49*, 114–137. https://doi.org/10.1111/anti.12185

Stone, M. E. (2006). What is housing affordability? The case of the residual income approach. *Housing Policy Debate, 17*(1), 151–184.

Thalmann, P. (2003). "House poor" or simply "poor"? *Journal of Housing Economics, 12*(4), 291–317.

Theodore, N., Peck, J., & Brenner, N. (2011). Neoliberal urbanism: Cities and the rules of markets. In G. Bridge & S. Watson (Eds.), *A new companion to the city. The city* (pp. 15–25). Wiley-Blackwell.

Theurillat, T., & Crevoisier, O. (2013). The Sustainability of financialized urban megaproject: The case of Sihlcity in Zurich. *International Journal of Urban and Regional Research, 37*(6), 2052–2073.

Theurillat, T., Rerat, P., & Crevoisier, O. (2014). The real estate markets: Players, institutions and territories. *Urban Studies, 52*(8), 1414–1433.

Touati-Morel, A. (2015). Hard and soft densification policies in the Paris city-region. *International Journal of Urban and Regional Research, 39*(3), 603–612.

Vallance, S., Perkins, H. C., & Dixon, J. E. (2011). What is social sustainability? A clarification of concepts. *Geoforum, 42*(3), 342–348. Elsevier Ltd.

Vollmer, L., & Kadi, J. (2018). Housing policy in the crisis of neoliberalism in Berlin and Vienna: Postneoliberal change of paradigm or selective pacification policy? *Prokla, 191*(48.2), 247–264.
Ward, C. (1974). *Tenants take over*. The Architectural Press Ltd.
WCED [World Commission on Environment and Development]. (1987). *Our common future*. United Nations.
Weingaertner, C., & Moberg, Å. (2014). Exploring social sustainability: Learning from perspectives on urban development and companies and products. *Sustainable Development, 22*(2), 122–133.
Williams, K. (2010). Sustainable cities: Research and practice challenges. *International Journal of Urban Sustainable Development, 1*, 128–133.
Woodcraft, S. (2012). Social sustainability and new communities: Moving from concept to practice in the UK. *Procedia—Social and Behavioral Sciences, 68*, 29–42.
Wyly, E., Newman, K., Schafran, A., & Lee, E. (2010). Displacing New York. *Environment and Planning A, 42*(11), 2602–2623.
Yiftachel, O., & Hedgcock, D. (1993). Urban social sustainability: The planning of an Australian city. *Cities, 10*, 139–157.
Zimmermann, E. W. (1933). *World resources and industries—A functional appraisial of the availability of agricultural and industrial resources*. Harper & Brothers Publishers.
Zukin, S. (1982). *Loft living: Culture and capital in urban change*. Johns Hopkins University Press.

CHAPTER 4

# Selecting Switzerland

## 4.1 Study Design and Methodology: Densification and Urban Housing Development in Switzerland

To explain the governance mechanisms at play leading to socially sustainable housing development in dense city areas (main research question), the data of the four articles of this book is conducted through a qualitative case study design. A case study is "an empirical inquiry that investigates a contemporary phenomenon in depth and within its real-life context, especially when the boundaries between the phenomenon and the context are not clearly evident" (Yin, 2018: 15). This qualitative approach makes it possible to uncover new trends and individuals' motivations in order to grasp a largely unknown and barely quantifiable process of social origin (George & Bennett, 2005; Taylor & Bogdan, 1998). It acknowledges that social sustainability in housing is produced by a complex interplay of contextual (institutional setting) and behavioral factors (actors' strategies). Housing development in dense urban environments cannot be separated from its context and thus requires a qualitative analysis (Flyvbjerg, 2006; Yin, 2018). By employing a qualitative case study design, the research questions (SQ1–3) are answered as results of human action and within its real-time socio-economic and -political context (Scholz & Tietje, 2002).

However, case study research also aims at generalization. It is connected to relational approaches in urban studies aiming to identify



G. Debrunner, *The Business of Densification*,
https://doi.org/10.1007/978-3-031-49014-9_4

similarities and/or differences in local patterns of policy implementation (Robinson, 2016; Ward, 2010). Deductively developed research variables—such as in this book, housing as a resource, institutions, and actors' strategies—postulate causal mechanisms (Sect. 3.7) that are investigated in the light of the empirical material collected in carefully selected cases (Sect. 4.3). Potential for generalization is not obtained through testing a hypothesis for general statistical significance but rather the results from the identified causal mechanisms, the relevance of which is expected to be broader than in the analyzed cases only (Flyvbjerg, 1998, 2006; Yin, 2018).

In the following section, I describe the Swiss context of the empirical analysis (4.2), the detailed reasons for case selection (4.3), as well as the different data collection and analysis methods applied (4.4). I conclude this chapter by reflecting upon my own positionality in the research process and the various challenges encountered in the field (4.6).

## 4.2 Study Context: Challenges of Housing, Urban Land Scarcity, and Social Exclusion in Switzerland

The present research is first and foremost based on a Swiss case study. The country counts 8.6 Mio. residents in total (FSO, 2020b) and is internationally considered a small state of only 41.285 square kilometers located in the center of Europe. Given the fact that much of the land cannot be or is not used for residential purposes due to landscape protection or mountainous surfaces, the effective population density is substantially higher than is average across Europe (Bourassa et al., 2010: 266). Consequently, Switzerland today has an urbanization level of about 73% (Weilenmann et al., 2017: 469).

The state makes a promising case study to gain knowledge on housing use conflicts as results of densification since the socio-economic challenges connected to urban growth have increased substantially in recent years (see Article 1 for details). Between 1935 and 2002, the degree of urban sprawl in Switzerland increased by 155%, and without effective policy measures, quantitative scenarios of future urban sprawl show that it is likely to further increase by more than 50% until 2050. Between 2002 and 2010, moreover, the degree of greenfield development was around three times as high as between 1980 and 2002 (Jaeger & Schwick, 2014: 294).

To reconcile concerns of urban sprawl, the Swiss federal government has introduced spatial development guidelines (the revised Federal Spatial Planning Act [SPA] in 2013) more in line with environmental sustainability in a span of more than 20 years (Swiss Federal Council, 2016). General principles include the restriction of urban sprawl, the reduction of energy emissions, and the support of a more compact urbanization (Rérat, 2012a: 116–120). Stated another way, densification has become a major issue in professional and planning circles as well as in the broader population, but there is still much questioning and concern that stands in the way of implementation at the municipal level (see Articles 2–4 for details) (Grams & Nebel, 2013; Nebel et al., 2017; Rérat, 2012a: 129; Swiss Federal Council, 2017). In sum, we are in the light of facing severe socio-spatial challenges linked to the end of greenfield development in this country.

Furthermore, the social implications of densification have become predominant in Swiss cities in recent years (Fig. 4.1). A general shift towards profit seeking in housing coupled with the obligation to densify introduced by the revised Federal Spatial Planning Act in 2013 have reinforced trends of social exclusion and gentrification in Swiss cities. In particular, old housing stocks are being demolished and redeveloped with higher rents (FOH, 2016a, 2016b; FOSI & FOH, 2015). As a consequence, a growing number of tenants living in urban rental housing stocks is confronted with eviction and displacement at short notice as they cannot afford the rents after densification and modernization tasks anymore (FOH, 2019: 4). Low- and middle-income households face difficulties in finding adequate housing as newly modernized apartments are primarily affordable for households with higher incomes, and non-profit housing suppliers have long waiting lists (Balmer & Gerber, 2017; FOH, 2017). Within the rental market, 28.9% of households suffer from excessive housing costs in relation to income (FOSI & FOH, 2015).

Simultaneously, population growth coupled with yield-oriented investments attracted by the state's economic stability and wealth has reinforced the attractiveness of Swiss housing markets. Triggered by low-interest rates, urban housing stocks have become the main target of capital investment, especially for pension funds (Theurillat & Crevoisier, 2013; Theurillat et al., 2014). As a result, social resistance strategies against densification and large-scale investment projects have increased in Swiss

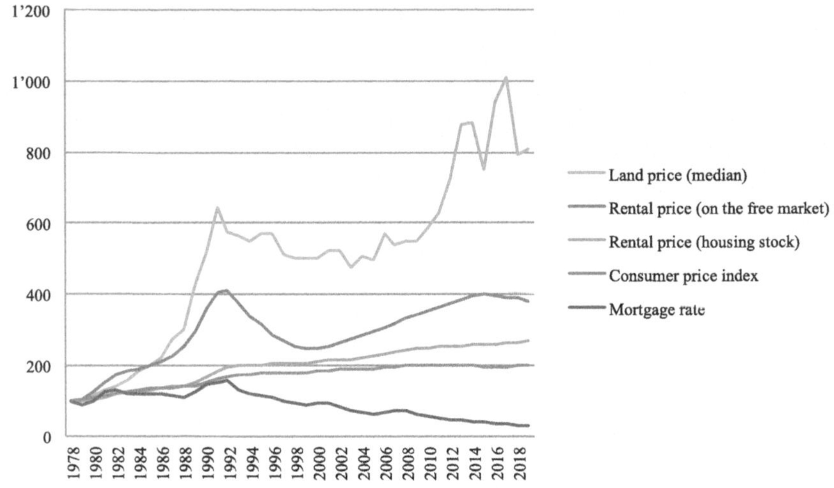


**Fig. 4.1** Housing development in Switzerland since 1978 (Index based: 1978 = 100). While mortgage rates have constantly decreased since 1978, prices for land and rents havesteadily risen (Canton of Zurich, 2020; FOH, 2020; FSO, 2020a, 2020c; Wüest & Partner, 2020)

cities because many tenants no longer accept the social implications caused by consolidation and upgrading (Maissen, 2018; Swiss Federal Council, 2017).

Due to these numerous reasons, Switzerland makes an interesting case study for the analysis of housing use conflicts and emerging socio-political challenges as results of densification. The federal state plays a crucial role in this matter in the sense that it signals how to deal with this issue also for cantons and municipalities. Switzerland is organized on three executive levels (municipalities, cantons, and the confederation) and characterized by a form of "cooperative federalism". Legislation in favor of densification goals and/or social sustainability of housing is introduced by the federal state and is to be implemented by cantons and municipalities (Linder, 1994).

Hence, the three institutional levels are jointly responsible for sustainable spatial development in Switzerland but have distinct areas of responsibility (principle of "subsidiarity"). The Swiss federal government is headed by the Federal Council (executive level), which is a collegial body consisting of seven ministers. They are elected by both chambers of parliament (legislative level) which consist of the National Council (representing the people) and the Council of States (representing the cantons). In general, the Swiss political system is characterized by direct democratic rights including the use of initiatives and referendums on all administrative levels (Bourassa et al., 2010: 268).

## 4.3 Case Selection: Housing Under Densification Pressure in Four Swiss Cities

To understand *how* housing is shaped in a context of densification, four Swiss case studies for detailed analysis were selected. The four cities—Zurich, Basel, Köniz, and Kloten—were chosen due to multiple reasons. *First*, the four municipalities were selected along the dependent variable of the main research question (social sustainability in housing). At the time of investigation, all four cities had to deal with ongoing housing challenges as results of urban densification projects. They all showed similar socio-economic characteristics such as strong population growth, scarcity of urban land, intensive densification pressure, affordable housing shortages, and challenges of new-built gentrification (Table 4.1). They moreover represented highly urbanized environments where most people live, where most economic development takes place, and where most housing projects are obtained. Hence, these cities are part of a highly integrated urban system in Switzerland in which the municipalities present regional centers for economic activities and living—similar to the urban structure in Germany or the Netherlands (Fig. 4.2). Therefore, by choosing these municipalities, the governance mechanisms at play (independent variable) leading to socially (un)sustainable housing development in a dense urban environment (main research question) became easily visible and graspable (Yin, 2018).

*Second*, while general densification objectives can be defined at a broader federal or regional scale, real-time implementation needs to be undertaken at the projects level, block by block, dealing with each impacted landowner one by one. To investigate concrete examples in the cases under scrutiny, I further selected cities that are currently challenged by concrete residential densification projects (opportunism). A (re)development area was defined as a set of buildings that belong to a private legal body and for which this body has a certain strategy to

**Table 4.1** Housing market characteristics in Zurich, Basel, Köniz and Kloten City (City of Kloten, 2020b; City of Köniz, 2020b; FOSD, 2017: 25; FSO, 2020b; Glattal Region, 2020; Nebel et al., 2017; Statistical Offices Zurich & Basel-City, 2019, 2020)

| | | | | | | *Housing property shares* | | |
|---|---|---|---|---|---|---|---|---|
| | *Urban context* | *Population in absolute numbers (2019)* | *Estimated population growth & demographic change* | *Housing vacancy rate (2019) (%)* | *Share of buildable lots within city boundaries (2017)* | *Owner-occupied (including condominium) (%)* | *Private rental housing (%)* | *Social rental housing (incl. public housing) (%)* |
| Switzerland | | 8603 Mio | +21% by 2050 | 1.66 | – | 37.1 | 58.6 | 4.3 |
| City of Zurich | *Core center* | 434,008 (around 1 Mio. including suburban areas) | +21% by 2030, significant increase of 1019 and 40–49 years old | 0.14 | 4% (for housing purposes only) | 47.2 | 28 | 24.9 |
| City of Basel | *Core center* | 178,445 (around 800,000 including suburban areas) | +10% by 2040, significant increase of children (10–19 years) and 4049 years old | 0.7 | 4% (only housing) | 49.7 | 36.1 | 13.5 |
| City of Köniz | *Suburban* | 42,694 | +6% by 2030 | 0.5 | 24% (only housing) | 15 | 75 | 10 |
| City of Kloten | *Suburban* | 20,079 | +5% by 2030 | 0.8 | 24% (only housing) | 20 | 71.3 | 8.7 |

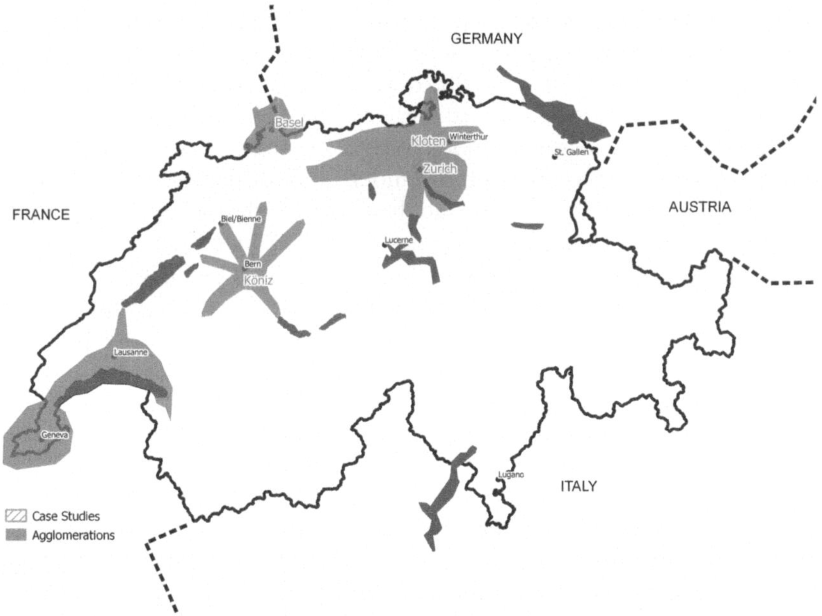


**Fig. 4.2** Map of Switzerland and the cases conducted within the country—Zurich, Basel, Köniz, and Kloten (own figure)

manage them (Fig. 4.2). In general, these densification areas were selected according to:

- The type of dominant use: **residential.**
- The type of landownership: **private rental.** Institutional investors own 63% of the housing stock property in Swiss cities. This ownership type is identified to be representative of many other residential housing areas in Swiss urban areas (FOH, 2017: 14).
- The project's actuality: **implemented within the last five years (2015–2020).** Thereby, I was able to directly confront the actors involved with their decisions and actions taken in order to understand *how* they cope with social sustainability challenges in housing under scarce land use conditions (see research questions).

*Third*, despite its similarities (e.g. housing challenges, densification pressure, population growth) the four cases also showed some distinct differences in terms of their local governance mechanisms (main research question). To compare and to assess the variation in their social, political, and institutional structures (independent variable), I selected municipalities in order to study and to understand differences in the local institutional context (SQ1) and applied actors' strategies (SQ2). Even though there are various public actors—at the national, regional, and local level—whose decision-making directly shapes and (re)defines the housing landscape, in liberal states such as Switzerland, the municipality is the public actor mainly responsible for addressing social challenges in housing (Rudolf et al., 2018: 476; Schönig, 2020). As regulators, municipal authorities (administrative level) on behalf of the city council (executive level), and the city parliament (legislative level) create, control, or dictate housing use rights. With respect to regulations on higher levels (e.g. Federal SPA), municipal authorities are in charge of granting building permits for residential densification projects to private landowners and building applications always need to align with the municipal zoning plan and its associated building ordinance (Bühlmann et al., 2011; Muggli, 2014).

I selected two urban core cities—the cities of Zurich and Basel (>100,000 inhabitants)—and two suburban cities—Köniz and Kloten (>20,000 inhabitants). While in core cities, authorities can usually rely on well-organized administrative units, smaller municipalities often have less capacity due to the lack of planning experts working in their administration. They are also confronted with a faster turnover among politicians, at both executive and legislative levels (Rudolf et al., 2018: 477). To understand *how* municipal governance works, I therefore selected four municipalities of different sizes and administrative structures to develop a broad understanding of the implementation of densification objectives in different urban contexts (Fig. 4.3).

### 4.3.1 The Case of Zurich City

The city of Zurich is a German-speaking city and the largest urban center in Switzerland (434,008 residents around 1 Mio. including the suburbs) (Statistical Office City of Zurich, 2020). The municipality of Zurich represents both the core center of Zurich agglomeration and the capital of the canton of Zurich. Due to its steady population and economic

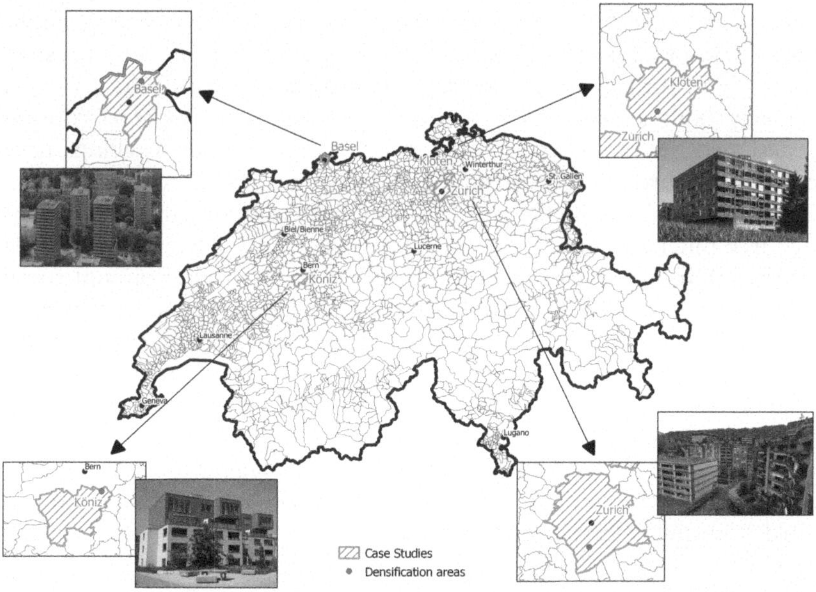


**Fig. 4.3** Densification areas within the four municipalities studied—Basel Schoren (top left), Kloten Southern district (top right), Köniz Nessleren (bottom left), and Zurich Brunau (bottom right) (own figure)

growth and its function as an international investment center, Zurich is considered Switzerland's financial capital (Theurillat & Crevoisier, 2013). It is moreover a metropolitan center characterized by a high degree of urbanity. For years, alongside Geneva, Zurich has been listed as one of the cities with the highest quality of life and at the same time the highest cost of living worldwide (Rérat & Lees, 2011: 131).

Since 1980, Zurich's population has increased by +17%, and investment in real estate has constantly risen (Statistical Office City of Zurich, 2020). During the 80s and 90s, Zurich started to deindustrialize and some of the abandoned industrial land was transformed into office development or was taken over for cultural activities. A major process of reconstruction began in which the municipality started to transform from "no-man's land" into a fashionable city. Many investors and developers have begun to recognize the city's potential and started to reclaim the buildings they owned (Rérat & Lees, 2011: 131).

Since the year 2000, however, the redevelopment of existing buildings or vacant plots has become increasingly challenging for investors due to conflicting land use interests at central locations. Only 10% of all newly built apartments have been built on unbuilt parcels during the last two decades in Zurich. Most of the newly built dwellings have been created through reconstruction and densification of existing housing stocks on already built land (e.g. through renovation, transformation) since free inner-city greenfield and brownfield areas are missing (City of Zurich, 2020). Therefore, the tensions between densification and social exclusion mechanisms have increased significantly in recent years (Rérat, 2012b). The absolute number of social evictions in the city's private rental sector has doubled within the period of 2006–2017 (Statistical Office City of Zurich, 2017). Between 2000 and 2013, rental prices in the housing stock have increased by 37% while rental prices on the free market have risen even more drastically, by 75% (Balmer & Gerber, 2017: 8). Moving to cheaper suburban areas has remained the main option to afford housing for many vulnerable and lower income groups living in Zurich.

#### *4.3.1.1 Zurich Brunau*

Research conducted in the city of Zurich focused on a residential area close to the center called "Zurich Brunau", a former clay pit and industrial zone (Figs. 4.4 and 4.5). Like many other parts of the city, the area was deindustrialized and transformed during the 1980s (Rérat, 2012a: 119). Since then, a series of new housing developments of a hundred units or more have been constructed. Specifically, Zurich Brunaupark—a settlement built in the 1980/1990s—is comprised of four residential buildings with 239 apartments and approximately 400 residents. It is situated next to the Sihlcity mega-project, which opened in 2007 as the first big urban entertainment center in Switzerland. Sihlcity is a shopping and leisure facility including almost 80 shops, nine cinema screens, a spa, a library, a hotel, a church, restaurants, several offices, and apartments (Theurillat & Crevoisier, 2013: 2062).

The pension fund of the bank "Credit Suisse" (CS) plans to demolish the inner-city location by replacing old buildings from the 1980s and 1990s. Four of the five buildings will be demolished and replaced by seven-storey residential buildings instead of today's five-storey residential buildings (potential for exploitation of +30,000 $m^2$). The number of apartments in the four new buildings will increase from 239 today

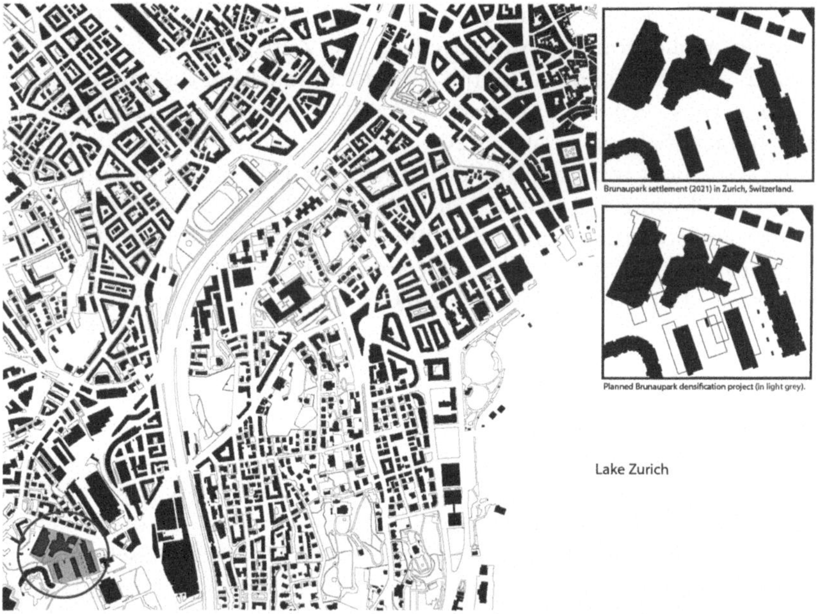


**Fig. 4.4** Map of Brunaupark settlement located at the South-Western part of the City of Zurich (top), and two pictures of the inner-yard of settlement itself (below) (*Sources* GIS data system, Canton of Zurich, 2022; own photographs)

**Fig. 4.5** Map of Brunaupark settlement located at the South-Western part of the City of Zurich (top), and two pictures of the inner-yard of settlement itself (below) (*Sources* GIS data system, Canton of Zurich, 2022; own photographs)

to 497 in the future (Schoop et al., 2020: 18). In March 2019, the planning application was submitted to the Zurich Office for Building Permits. This was followed by the contract termination for a total of 239 tenant parties. Within a year, around 450 people would have to leave their homes (Schoop et al., 2020). In 2012, two of the four buildings in the settlement were refurbished, namely, kitchens and bathrooms were redone. Following the planned demolition and new construction, rents are expected to rise by around +60%. According to the investor, a 3.5 room apartment (75 $m^2$) will cost between 2200 and 2650 CHF monthly net rent (today about 1500 CHF net).

### *4.3.2 The Case of Basel-City*

The city of Basel is a German-speaking city and after Zurich and Geneva the third largest urban center in Switzerland (178,445 inhabitants; around 800,000 including suburban areas) (Swiss Cities Association, 2020). Basel-City is also the capital of the canton of Basel-Stadt, which it forms with the municipalities of Riehen and Bettingen. Basel is moreover a border town located on the Northern border of Switzerland at the triangle between Switzerland, Germany, and France. The city therefore has suburbs in all three countries. Basel is divided into "Grossbasel" on the left (Southwestern) side of the Rhine River and the area of "Kleinbasel" on the right (Northeastern) riverbank of the Rhine. The municipality is considered a global center for the chemical and pharmaceutical industry. Two large international pharmaceutical companies "Novartis" and "Roche" have their headquarters in Basel.

From a spatial development perspective, Basel-City has undergone a period of structural change and intensive population growth since the 1990s (Ott, 2020). Since 2000, the city population has increased by +7% to 178,445 in total (Statistical Office City of Basel, 2020). By 2040, Basel's residential population is expected to increase by another 10%. The city currently faces a shortage of affordable housing. Between 2003 and 2013, rents in existing stocks have increased by 29%. Rental prices offered on the free market have increased even more dramatically, by 38% (Balmer & Gerber, 2017: 8). An increasing number of socially disadvantaged groups find it difficult to find an apartment within city boundaries (Basel-City Council, 2018: 6). To address these challenges, the city government has initiated the "1000+" housing program, which aims to create a total of 1000 newly built affordable apartments in public ownership by 2035. In addition, many former industrial sites have been redeveloped or are currently being transformed into housing areas (Ott, 2020; Statistical Office City of Basel, 2019: 6).

#### *4.3.2.1 Basel Schoren*

The empirical analysis conducted in the city of Basel focused on a residential area close to "Basel Badischer Bahnhof", situated at the Northeastern edge of the city center. The "Schoren" area includes some newly built developments of different sizes on former brownfield sites or on vacant land, but also the redevelopment of already densified land. Around 1400 inhabitants currently live in the Schoren neighborhood, and the population is expected to grow steadily in the next few years. In particular, the relocation of the former Novartis office campus has created vacant industrial space that has now been designated to be transformed into a large-scale residential area. From 2012 to 2018, an additional 800 residents moved to the Schoren area, which meant a population increase of almost +30% (City of Basel, 2015: 2). Specifically, I focused on the Basel Schorenweg area—a settlement built in 1961—which counts 196 apartments with around 300 residents in total. The Credit Suisse investment fund plans to densify the two buildings via total internal reconstruction with smaller housing units in 2021 (Laur, 2019: 21) (Figs. 4.6 and 4.7.

### *4.3.3 The Case of Köniz City*

Köniz is a German-speaking municipality in the "Bern-Mittelland" administrative district in the canton of Bern in Switzerland. The municipality is part of the wider agglomeration of the city of Bern. Köniz is located southwest of Bern center and is the fourth largest municipality in the canton of Bern and the thirteenth largest town in Switzerland with 42,694 inhabitants in total (City of Köniz, 2020a). Between 2003 and 2013, the residential population in Köniz increased by 7%, which is more than in the surrounding municipalities. The increase has primarily to do with the proximity of Köniz to Bern and immigration from abroad. In principle, the majority of the households living in Köniz can afford an apartment on the rental market. However, it is assumed that, due to high demand and scarce land use conditions in the municipality, the housing situation will worsen in the future. Prices on the housing market tend to rise, which primarily affects family households and lower income groups (Beck et al., 2016: 14–15).

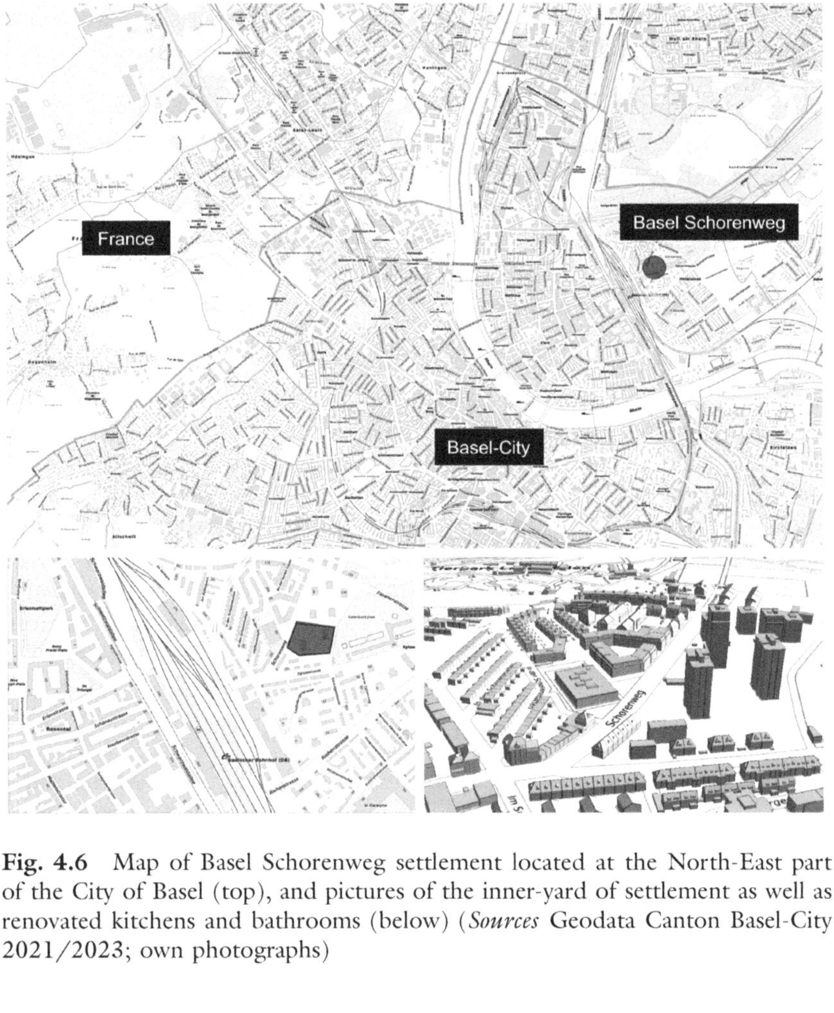


**Fig. 4.6** Map of Basel Schorenweg settlement located at the North-East part of the City of Basel (top), and pictures of the inner-yard of settlement as well as renovated kitchens and bathrooms (below) (*Sources* Geodata Canton Basel-City 2021/2023; own photographs)

#### *4.3.3.1 Köniz Nessleren*

Research conducted in the city of Köniz focused on the "Nessleren" area located in the Wabern district at the northeastern edge of the municipality. The settlement consists of 33 houses that were built between 1979 and 1982. The three-story buildings are arranged in 13 rows of two or three houses. The three institutional owners of the settlement (one private foundation, one private bank, and one private insurance company)

**Fig. 4.7** Map of Basel Schorenweg settlement located at the North-East part of the City of Basel (top), and pictures of the inner-yard of settlement as well as renovated kitchens and bathrooms (below) (*Sources* Geodata Canton Basel-City 2021/2023; own photographs)

**Fig. 4.8** Map of Köniz Nessleren settlement located at the Eastern part of the City of Köniz (top), and pictures of the former (left) and newly transformed and densified housing stock (right) (*Sources* Geodata Canton Bern 2021/2023; own photographs)

decided to remedy the existing structural deficits, in particular, the insulation and heating system through reconstruction. The buildings were completely renovated and densified in 2018. The settlement now counts 60% additional residential units (Espace Suisse, 2018) (Figs. 4.8 and 4.9).

### *4.3.4 The Case of Kloten City*

The city of Kloten is a German-speaking city located about 10 km northeast of Zurich city center.

**Fig. 4.9** Map of Köniz Nessleren settlement located at the Eastern part of the City of Köniz (top), and pictures of the former (left) and newly transformed and densified housing stock (right) (*Sources* Geodata Canton Bern 2021/2023; own photographs)

The municipality is part of Zurich's agglomeration and metropolitan area (AZMA, 2020). From 1946 to 1948, Zurich-Kloten airport was built west of the village of Kloten. The city is also close to the airport motorway, which connects the town to the (inter)national highway system (City of Kloten, 2020a). Since 1980, the city's population has increased by 27%. It is estimated that Kloten will increase by another 15% by 2040 (Glattal Region, 2020). By 2020, the "Circle Project" will open at Zurich airport, which will create around 4000 new jobs in the city. This population growth is difficult to combat within Kloten's municipal boundaries. The city does not have any free unbuilt reserve zones left and therefore must densify via soft measures (e.g. conversion). Most of the buildings were built in the 1960s and 1970s and need modernization. In addition, the city of Kloten with residents from around 120 different nations is confronted with a very high degree of residential fluctuation. About 50% of the population leaves Kloten within five years. The main reason for this is the airport: many international residents have temporary positions and are only in the country for a short period. Rents are affordable for most residents living in Kloten; however, confronted with the situation that old housing stocks are currently being demolished and densified, many tenants struggle with social displacement because the upgrades lead to higher rents. In some city areas, a clear process of social segregation is occurring where particularly lower income groups can no longer afford housing (City of Kloten, 2019: 1–12).

#### *4.3.4.1 Kloten Southern District*

Empirical work conducted in the city of Kloten focused on the residential area at the southern edge of the city next to the municipal border to "Opfikon-Glattbrugg". Due to the representative location at the entrance of the municipality, the city council decided to improve the urban situation in this area. It is planned that the district will transform into an attractive urban area in the next few years. Many new workplaces at Zurich Airport have been created, which is why an increasing number of residents are expected to move to the Kloten Southern district (City of Kloten, 2019). The area is comprised of around 20 plots owned by private institutional investors and individuals. The "kloten. milano" project—the settlement I investigated—was demolished and rebuilt with triple use density in 2016. The owner is an institutional investor and developer from the Zurich region. The around 80 residents living in the former settlement had to leave their apartments in 2016 as most of them could not afford the higher rents in the modernized buildings anymore (City of Kloten, 2015) (Figs. 4.10 and 4.11).

## 4.4 Data Collection Methods

The empirical material of this book was conducted by the use of qualitative research methods. This was done in order to gain a detailed understanding of the governance mechanisms at play (institutional rules, actors' strategies) leading to socially sustainable housing development (see main research questions) (Flick, 2007; Gläser & Laudel, 2010). More precisely, the three deductively developed key variables of the theoretical model—housing as a resource, institutions, and actors' strategies—were analyzed by employing qualitative methods that facilitate gathering information of social origin (Yin, 2018). All data collection methods employed were intended to aid a better understanding of the complex relationships between housing and densification dynamics in order to answer the research questions. Specifically, the following research methods were employed: qualitative document analysis, participant observation, household surveys, semi-structured and expert interviews. The actual field research was carried out between August and September 2018 (Articles 1 and 4) and between May and October 2019 (Articles 2 and 3). A comprehensive list of all interviews is provided in Annex 1. The research process

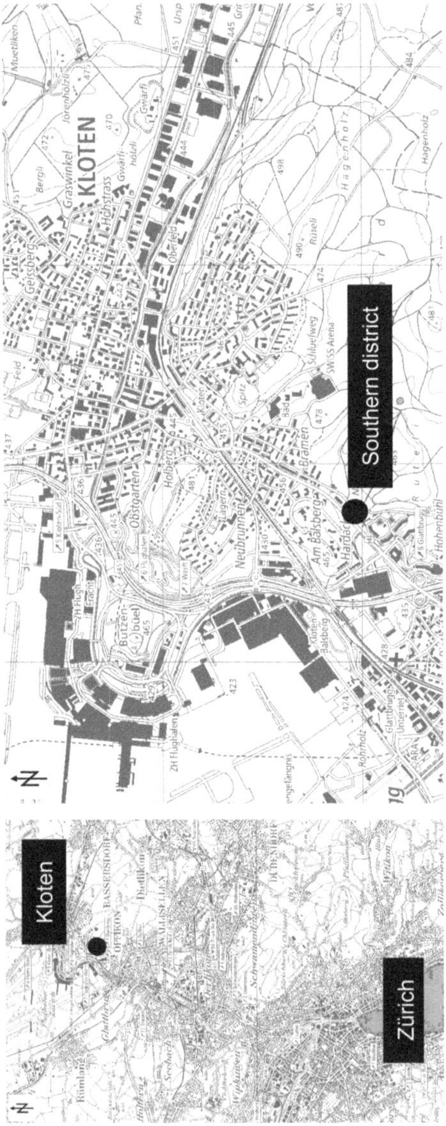


**Fig. 4.10** Map of Kloten Southern district (*Source* GIS Browser Canton of Zurich, 2023)

**Fig. 4.11** Photographs of Kloten Southern district settlement before (top) and after (below) demolishment and total replacement construction and densification (*Sources* Google Streetview; own photographs)

was moreover organized circularly. The phases of data collection, analysis, and interpretation were not necessarily regarded as conceptually distinct but as interconnected throughout the process (Behnke et al., 2010: 42).

### *4.4.1 Document and Statistical Data Analysis*

Since this research is concerned with housing, the *firststep* of data collection before going to the field was a statistical data analysis (but qualitative in nature) focusing in particular on the socio-economic state of housing at the federal level as well as in each municipality (population dynamics, development of rents, vacancy rates, land, and housing prices, etc.). The analysis helped me to capture housing (re)development within its real-life socio-economic context and concerning potential use conflicts. Moreover, the housing situation in each city was related to the general housing situation in Switzerland (George & Bennett, 2005). This step helped me to further elaborate upon the dependent variable of the main research question (social sustainability in housing).

In a *second step*, I analyzed the regulatory institutions of the housing stock both at the federal, cantonal, and municipal levels (sub-question 1). This helped me to understand the institutional regime governing housing under densification pressure. Since it was impossible to review the full extent of laws and policies that constitute the institutional regime, I primarily focused on regulations and policy instruments that were considered to have the most effect on the housing stock's sustainability. Besides housing, planning, and building laws directly impacting the extent and range of housing, these also included more indirect ways of public intervention such as environmental, monument protection, tax laws as well as property rights, contracts, and tenancy law. The aim of the content analysis was to structurally filter the documents in relation to certain topics and aspects of the problem of interest and to summarize them (Mayring, 2010).

In all four papers, I performed a broad screening of policy documents at the interface between urban densification, housing, and social sustainability issues (affordable housing, social mixing, etc.). The qualitative analysis included written sources such as government reports, legislation, strategy papers, and parliamentary debates that were primarily published within the last decade. I also incorporated newspaper articles, project documents, and "grey literature" in order to understand the characteristics of the specific formal rules in force.

### *4.4.2 Participant Observation*

A second important source of information was provided by participatory observations (Reuber & Pfaffenbach, 2005). I employed this step in order to better understand the research context and to gain knowledge *prior* to the interviews in which actors' strategies and objectives were assessed (sub-question 2). To do so, I spent a lot of time visiting the cities and densification areas selected and sought to speak to as many people as possible in order to gain various insights from residents and other stakeholders. For instance, in Zurich Brunau, I spent a considerable amount of time at the playground and in the backyard speaking to parents or elderly people who had only recently received their contract termination. A variety of people used to come to have a chat with me as they noticed that I was taking notes and documenting my impressions when visiting the place.

In Basel Schorenweg, I visited an 86-year-old female resident, who was living alone in her flat struggling to find an alternative dwelling. A nurse who helped her to cope with the situation supported her. I spoke to both several times in order to grasp valuable insights into how older people deal with densification challenges. While I was mostly just an observer on these occasions, I also participated in tenants group meetings led by the local tenants' associations. These conservations revealed a lot about life in the densification area, about neighborhood relations or conflicts as well as investors' and public authorities' communication strategies with tenants.

Furthermore, I attended opening celebrations of newly finished densification projects to gather background information on how investors communicate and how their internal organizational structures work. Thereby, I also gained knowledge of how the relations between the investor, the developer, and other supplying firms of the real-estate industry such as private architecture or planning offices are managed. During my doctoral studies, I also attended a six-month program of advanced studies in urban management at the University of Zurich. In this course, I met professionals working for the private real-estate industry and learned a lot about their decision-making behavior. These gatherings granted me important insights into the discourses of economic players that they use to legitimize their actions.

During both research periods, I constantly wrote down my analytical thoughts and memos in my field book (Charmaz, 2008: 162). I noted all observations and informal conversations and tried to record as many details as possible right after the informal discussions. I wrote down specific questions for which I ought to seek out specific respondents to constantly improve my knowledge and to precisely tailor my interview questions.

### 4.4.3 *Household Surveys*

As outlined in the theory part of this book, the principal source of evidence concerning socially sustainable housing development in densification processes is considered to be people themselves, particularly those living in the areas in question (Bramley et al., 2009: 2129). With this in mind, I conducted a household survey with 412 households living in the settlements of Zurich Brunau and Basel Schoren to gain a detailed understanding of the residents' perspectives (sub-question 2). The survey enabled me to determine the profile of the residents, their motivations, and the socio-economic challenges they are currently confronted with. The two large-scale densification areas were selected for detailed comparative analysis of households' perceptions as they are both owned by the same institutional investor (Credit Suisse [CS]; see Article 2). Moreover, CS' projects in Zurich and Basel were both ongoing at the time of investigation (between May and October 2019), which is why the tenants involved could be directly confronted with their decisions and actions taken. By following this project-based approach, I investigated social sustainability in housing "from the ground up, as it actually exists in local places, and as a set of evolving practices" (Krueger & Agyeman, 2005: 416).

The household survey incorporated the social sustainability indicators presented in Article 2 and included both open and multiple-choice questions. The open questions were used to gain a rich understanding of the households' perspectives on how tenants are affected by densification plans. The multiple-choice questions were used to further underline household positions, but the analysis remained qualitative in nature. I opted for a self-completion postal and digital survey method (with one reminder) and managed to achieve a respectable 25% response rate (101 responses in total). In designing the questionnaire, I considered the existing body of literature as well as a number of national surveys covering similar topics that helped me to identify whether and how questions have been shown to work.

### 4.4.4 Semi-Structured and Expert Interviews

To understand the diverse strategies and the behavior of the actors involved in densification projects (sub-question 2), I employed 54 semi-structured interviews with representatives from both the public and the private sector. These included policymakers from the national to the local district level, practitioners from public ministries, and representatives of homeowners' and tenants' associations, housing cooperatives, neighbors, and residents (Annex 1).

Out of these, 42 were conducted as expert interviews and 12 as semi-structured interviews with residents. My intention in performing expert interviews was to gain detailed information about how specific individuals and activist groups perceive densification in relation to their particular function (Meuser & Nagel, 2009: 57). Experts interviews are particularly useful in research settings that intend to identify causal mechanisms that are to be analyzed in a more detailed manner and from a range of different perspectives (Blatter et al., 2007: 60). For example, I interviewed several politicians due to their expert knowledge and professional position in parliament or in public administration. By conducting interviews, my goal was to investigate the actors' reasons and motivation to participate in residential densification (SQ2). All experts were chosen due to their detailed understanding and knowledge of the topic as well as based on their practical expertise related to the position they occupied within certain institutional structures.

For the interviews, I prepared thematically structured guidelines in the format of a semi-structured questionnaire. I sent the guiding questions to the interviewees in advance so that they could prepare for discussion. In contrast to standardized interviews, performing semi-structured interviews enabled me to explore the interviewees' knowledge during the interview process in an explorative manner (Gläser & Laudel, 2010). All interviews were conducted in person and were mostly held in the offices of the respondents or alternatively at a location they chose (e.g. one politician was interviewed in a restaurant). All interviews were recorded with the permission of the respondents. In the majority of cases, interview participants were alone, and the interviews could proceed undisturbed. All interviews lasted between 60 and 90 minutes. I stopped interviewing people when no further knowledge could be gathered, or the same information was repeated by different sources.

## 4.5 Data Analysis Methods

The qualitative data material—either in the format of documents, surveys, or interviews—was analyzed by following general principles of qualitative content analysis (Gläser & Laudel, 2010: 46; Mayring, 2010). The three deductive variables—*housing as a resource*, *institutions*, and *actors' strategies*—and the mechanisms that bind their interaction (independent variable) were identified in the text via a code-based context analysis. First, the non-written data material (audio interviews) was transcribed into text using a professional transcription service and then coded with the help of MaxQDA as the data analysis tool. The software aided in deleting or rewriting specific codes in an effective manner. The text was then analyzed along the three variables at play and related to the specific themes the interviewees raised in each of the three subsections. I however did not only focus on deductive coding. Rather I combined inductive and deductive coding and remained open to emergent themes during the process of data analysis (Charmaz, 2008; Emerson et al., 1995; Glaser & Strauss, 1967). In essence, I coded my data using thematic codes based on my research questions and theoretical concepts as well as more analytical codes that inductively emerged from the data.

## 4.6 Positionality, Reflexivity, and Validity of the Data

The variety of data collection methods as well as the time spent in the municipalities and densification areas strengthened the validity and reliability of the results. Risks of selective data acquisition were reduced by triangulating different qualitative methods. However, every method applied has its own limitations that I would like to briefly reflect on to arrange the results within an appropriate framework.

The questionnaire conducted with tenants is based on the self-assessment of the respondents under the basic assumption that participants respond to the surveys according to their best knowledge and based on subjective perspectives. The qualitative statements, however, were not verified or compared on a more general basis. Even though the variables incorporated in the questionnaire were justified by referring to theory, there is still a possibility that relevant data could not be effectively captured. The statements made in the questionnaire helped to interpret some of the results; however, a closer examination (e.g. in-depth expert interviews or broader statistical analyses) is imperative for more valid derivations.

When performing interviews, the validity of the data was increased by sending the questions to the participants in advance so that they could prepare for the task. In addition, training was carried out with the participants at the beginning of each interview in order to make the data collection process more transparent. I also paid particular attention to data documentation (e.g. in transcription, and field notes) to make each step of data collection explicit and replicable (Gläser & Laudel, 2010: 193). To guarantee data protection, I moreover asked all residents interviewed to sign a document for ethical approval to ensure that the data collected in their homes can—in an anonymous way—be used for publication.

Finally, my identity as a researcher played a key role in the data collection process and the interactions with various stakeholders in the field. I noticed this in particular in discussions with institutional investors or politicians who, at the beginning, were very skeptical towards land policy research since they obviously perceived it as a left-wing policy. While trying to analyze densification from a social science perspective, I became aware that I am simultaneously an actor. Urban science itself is a social area with its own power games. My scientific knowledge is always subject to strong uncertainty, positionality, but also subjectivity. Since I identify myself as a critical human geographer, it was very important for me to explain the background of my work in detail *before* gathering data in order to create mutual trust and an open atmosphere. I made my research objectives very clear from the start and always explained my interests in *analyzing* (not judging) densification processes from a human geography and political ecology perspective, which includes a focus on power games.

Furthermore, the sensitivity to the tenants' housing situation was difficult to manage in many ways. It was never my intention to encroach upon their privacy, but to collect valuable insights I sometimes had to ask very personal questions, for example, concerning their future housing options even though they only recently received contract termination. In these situations, it was my impression, however, that my age or gender (or both) helped to establish a connection with them. They considered me more of a friend than a pure researcher and were very open-minded to tell me about their current living situation. After all, keeping a research diary was an important part of reflecting on my positionality and the way it affected my interactions with different stakeholders involved in the process.

## References

AZMA [Association Zurich Metropolitan Area]. (2020). *Entwicklung Metropolitanraum Zürich*. Available at https://www.metropolitanraum-zuerich.ch/home.html. Accessed 18 October 2020.

Balmer, I., & Gerber, J.-D. (2017). 'Why are housing cooperatives successful? Insights from Swiss affordable housing policy. *Housing Studies, 33*, 361–385.

Basel-City Council. (2018). *Controlling Bericht Wohnraumentwicklung 2012–2017*.

Beck, L., Fuchs, S., & Thomas, M. (2016). *Grundlagenbericht zur Initiative "bezahlbar Wohnen in Köniz"*.

Behnke, J., Baur, N., & Behnke, N. (2010). *Empirische Methoden der Politikwissenschaft*. UTB. Schöningh.

Blatter, J., Janning, F., & Wagemann, C. (2007). *Qualitative Politikanalyse, Grundwissen Politik*. VS Verlag für Sozialwissenschaften.
Bourassa, S., Hoesli, C., & Scognamiglio, D. (2010). Housing finance, prices, and tenure in Switzerland. *Journal of Real Estate Literature, 18*(2), 262–282.
Bramley, G., Dempsey, N., Power, S., Brown, C., & Watkins, D. (2009). Social sustainability and urban form: Evidence from five British cities. *Environment and Planning A, 41*(9), 2125–2142.
Bühlmann, L., Haag, H., Jud, B., Kissling, S., & Spori, N. (2011). Einführung in die Raumplanung [Introduction to Land Use Planning]. VLP-ASPAN *Raum&Umwelt*.
Canton of Zurich. (2020). *Entwicklung der Landpreise Kanton Zürich, Entwicklung der Landpreise im Kanton Zürich*. Available at: https://www.zh.ch/de/planen-bauen/raumplanung/immobilienmarkt/bodenpreise.html. Accessed 18 October 2020.
Canton of Zurich. (2023). *GIS Browser online*. Accessed: 13 Dcember 2023.
Charmaz, K. (2008). Grounded Theory as an emergent method. In S. N. Hesse-Biber & P. Leavy (Eds.), *Handbook of emergent methods* (pp. 155–172). The Guilford Press.
City of Basel. (2015). *Quartierarbeit Hirzbrunnen*.
City of Kloten. (2015). *Protokoll Gemeinderat Kloten*.
City of Kloten. (2019). *Protokoll Stadtrat Kloten: Klotener Volksinitiative "wohnen für alle": Antrag auf Gültigkeitserklärung und Ablehnung der Initiative*.
City of Kloten. (2020a). *Entwicklung Stadt Kloten*. Available at www.kloten.ch. Accessed 18 October 2020.
City of Kloten. (2020b). *Report on the local housing situation, municipal planning authority on behalf of the city council* (pp. 1–12).
City of Köniz. (2020a). *Gemeinde Köniz in Zahlen*. Available at https://www.koeniz.ch. Accessed 18 October 2020.
City of Köniz. (2020b). Statistics and socio-economic development. Available at https://www.koeniz.ch/wohnen/zuzug-umzug-wegzug/zuzug-umzug-wegzug.page/237. Accessed 7 July 2020.
City of Zurich. (2020). *Bauliche Verdichtung aktuell*. Available at https://www.stadt-zuerich.ch/content/prd/de/index/statistik/publikationen-angebote/publikationen/webartikel/2020-04-07_Bauliche-Verdichtung-aktuell.html. Accessed 18 October 2020.
Emerson, R. M., Fretz, R. I., & Shaw, L. L. (1995). *Writingethnographicfieldnotes*. University of Chicago Press.
Espace Suisse. (2018). *Im Ortsteil Wabern wurde die Siedlung Nessleren vertikal verdichtet*.
Flick, U. (2007). *Qualitative Sozialforschung*. Rohwolt Taschenbuch Verlag.

Flyvbjerg, B. (1998). *Rationality and power democracy in practice*. University of Chicago Press.

Flyvbjerg, B. (2006). Five misunderstandings about case-study research. *Qualitative Inquiry, 12*(2), 219–245.

FOH [Federal Office for Housing]. (2016a). *Bericht Energetische Gebäudesanierungen im Mietwohnungsbereich*.

FOH. (2016b). *Wohnungspolitischer Dialog Bund, Kantone und Städte Zwischenbericht der Arbeitsgruppe*.

FOH. (2017). *Gemeinnütziges Wohnen im Fokus. Ein Vergleich zu Miete und Eigentum*.

FOH. (2019). *Zusammenhang zwischen Einkommens- und Energiearmut sowie die Folgen energetischer Sanierungen für vulnerable Gruppen. Eine qualitative Analyse*.

FOH. (2020). *Hypothekarzinsentwicklung, Entwicklung Referenzzinssatz*. Available at https://www.bwo.admin.ch/bwo/de/home/mietrecht/referenzzinssatz.html. Accessed 18 October 2020.

FOSD [Federal Office for Spatial Development]. (2017). *Bauzonenstatistik Schweiz 2017. Statistik und Analysen*. Available at https://www.are.admin.ch/dam/are/de/dokumente/grundlagen/dokumente/bauzonenstatistik-schweiz-2017-statistik-und-analysen.pdf.download.pdf/Bauzonenstatistik_Schweiz_2017.pdf. Accessed 18 October 2020.

FOSI [Federal Office for Social Insurances] and FOH [Federal Office for Housing]. (2015). *Wohnungsversorgung in der Schweiz—Bestandesaufnahme über Haushalte von Menschen in Armut und prekären Lebenslagen*.

FSO [Federal Statistical Office]. (2015). *Bevölkerunsstand- und Struktur*.

FSO [Federal Statistical Office]. (2020a). *Bestandsmietentwicklung*. Available at https://www.bfs.admin.ch/bfs/de/home/statistiken/preise/mieten/index.html. Accessed 18 October 2020.

FSO [Federal Statistical Office]. (2020b). *Bevölkerungsentwicklung von 2020 bis 2050: Wachstum, Alterung und Konzentration rund im die grossen Städte*.

FSO [Federal Statistical Office]. (2020c). *Index der Konsumentenpreise, Entwicklung Konsumentenpreise*.

George, A. L., & Bennett, A. (2005). *Case studies and theory development in the social sciences*. MIT Press.

Glaser, B. G., & Strauss, A. L. (1967). *The discovery of grounded theory*. Aldine.

Gläser, J., & Laudel, G. (2010). *Experteninterviews und qualitative inhaltsanalyse* (4th ed.). VS Verlag für Sozialwissenschaften.

Glattal Region. (2020). *Portrait Region Glattal, Bevölkerungsentwicklung*. Available at https://zpg.ch/portrait. Accessed 7 July 2020.

Grams, A., & Nebel, R. (2013). Nutzungsreserven und -potenziale für die Siedlungsentwicklung nach innen. *disP—The Planning Review, 49*(2), 28–38.

Jaeger, J. A. G., & Schwick, C. (2014). Improving the measurement of urban sprawl: Weighted Urban Proliferation (WUP) and its application to Switzerland. *Ecological Indicators, 38*, 294–308.

Krueger, R., & Agyeman, J. (2005). Sustainability schizophrenia or "actually existing sustainabilities?" Toward a broader understanding of the politics and promise of local sustainability in the US. *Geoforum, 36*(4), 410–417.

Laur, F. (2019, August 27). Das Geisterhaus am Schorenweg. *Basler Zeitung*, p. 21.

Linder, W. (1994). *Swiss democracy: Possible solutions to conflict in multicultural societies*. St. Martin's Press.

Maissen, F. (2018). Rendite um jeden Preis? Massenkündigungen von Mietwohnungen nehmen zu. *Swiss Radio and Television*. Available at https://www.srf.ch/news/schweiz/rendite-um-jeden-preis-massenkuendigungen-von-mietwohnungen-nehmen-zu. Accessed 22 July 2019.

Mayring, P. (2010) *Qualitative Inhaltsanalyse. Grundlagen und Techniken*. 11. Auflag. Beltz Verlag.

Meuser, M., & Nagel, U. (2009). Experteninterview und der Wandel der Wissensproduktion Wiesbaden. In A. Bogner (Ed.), *Experteninterviews. Theorie, Methoden, Anwendungsfelder. Verlag für Sozialwissenschaften* (pp. 35–60).

Muggli, R. (2014). *Spatial planning in Switzerland: A short introduction*. VLP-ASPAN *Raum & Umwelt*.

Nebel, R., Hollenstein, K., Di Carlo, G., Niedermaier, M., & Scholl, B. (2017). *Schweizweite Abschätzung der Nutzungsreserven 2017*. ETH University.

Ott, L. (2020). *Anfrage zur Mithilfe bei der Beantwortung der Kommunikation der UNO-Sonderberichterstatterin für angemessenen Wohnraum*.

Rérat, P., & Lees, L. (2011). Spatial capital, gentrification and mobility: Evidence from Swiss core cities. *Transactions of the Institute of British Geographers, 36*(1), 126–142.

Rérat, P. (2012a). The new demographic growth of cities: The case of reurbanisation in Switzerland. *Urban Studies, 49*, 1107–1125.

Rérat, P. (2012b). Housing, the compact city and sustainable development: Some insights from recent urban trends in Switzerland. *International Journal of Housing Policy, 12*(2), 115–136.

Reuber, P., & Pfaffenbach, C. (2005). *Das Geographische Seminar: Methoden der empirischen Humangeographie*. 1. Auflage. Westermann.

Robinson, J. (2016). Comparative urbanism: New geographies and cultures of theorizing the urban. *International Journal of Urban and Regional Research, 40*(1), 187–199.

Rudolf, S. C., Kienast, F., & Hersperger, A. M. (2018). Planning for compact urban forms: Local growth-management approaches and their evolution over time. *Journal of Environmental Planning and Management, 61*(3), 474–492.

Scholz, R., & Tietje, O. (2002). *Embedded case study methods: Integrating quantitative and qualitative knowledge*. Sage.

Schönig, B. (2020). Paradigm shifts in social housing after welfare-state transformation: Learning from the German experience. *International Journal of Urban and Regional Research*, 1–18. https://doi.org/10.1111/1468-2427.12914

Schoop, F., Kälin, A., & Hofer, K. (2020). Das lange Warten auf den Abbruch. *Neue Zürcher Zeitung*, pp. 18–19.

Statistics Basel-City. (2019). *Wanderungsbefragung 2018. Ergebnisbericht*.

Statistical Office City of Basel. (2019). *Bevoelkerungsszenarien*.

Statistical Office City of Basel. (2020). *Bevölkerungsentwicklung Stadt Basel, Bevölkerungsentwicklung*. Available at https://www.statistik.bs.ch. Accessed 18 October 2020.

Statistical Office City of Zurich. (2017). *Immer mehr Leerkündigungen bei Umbauten*. Available at https://www.stadt-zuerich.ch/prd/de/index/statistik/publikationen-angebote/publikationen/webartikel/2017-09-14_Immer-mehr-Kuendigungen-bei-Umbauten.html. Accessed 26 March 2020.

Statistical Office City of Zurich. (2020). *Bevölkerungs- und Wirtschaftsentwicklung Stadt Zürich*. Available at https://www.stadt-zuerich.ch/prd/de/index/statistik/themen/bevoelkerung.html. Accessed 25 June 2020.

Swiss Cities Association. (2020). *Statistik der Schweizer Städte 2020*.

Swiss Federal Council. (2002). *Strategie Nachhaltige Entwicklung 2002*.

Swiss Federal Council. (2016). *15.044 Geschäft des Bundesrates zur Änderung des OR. Mietrecht*.

Swiss Federal Council. (2017). *Verdichtetes Bauen in Ortszentren fördern, aber wie? Bericht des Bundesrates in Erfüllung des Postulats von Graffenried 14.3806 vom 24. September 2014*.

Taylor, S. J., & Bogdan, R. (1998). *Introduction to qualitative research methods. A guidebook and resource*. Wiley.

Theurillat, T., & Crevoisier, O. (2013). The sustainability of financialized urban megaproject: The case of sihlcity in Zurich. *International Journal of Urban and Regional Research, 37*(6), 2052–2073.

Theurillat, T., Rerat, P., & Crevoisier, O. (2014). The real estate markets: Players, institutions and territories. *Urban Studies, 52*(8), 1414–1433.

Ward, K. (2010). Towards a relational comparative approach to the study of cities. *Progress in Human Geography, 34*(4), 471–487.

Weilenmann, B., Seidl, I., & Schulz, T. (2017). The socio-economic determinants of urban sprawl between 1980 and 2010 in Switzerland. *Landscape and Urban Planning, 157*, 468–482.

Wüest & Partner. (2020). *Angebotsmieten: Analyse der jährlichen Angebotspreisentwicklung*.

Yin, R. K. (2018). *Case study research and applications. Design and methods* (6th ed.). Sage.

CHAPTER 5

# Investigating Switzerland

## 5.1 Investigating Switzerland: Learning from the Swiss Scarce Land Use Situation

The empirical analyses in the four articles that compose this book focus on social sustainability challenges of housing in a context of densification in Swiss cities. All four articles examine *how* densification leads to changing housing use situations in the Swiss urban rental sector, and *how* conflicting use interests between competing actors at different geographical scales evolve, and can potentially be solved. The theoretically determined variables—*social sustainability in housing (condition of the housing resource)* (Sect. 3.2), *institutions* (Sect. 3.4), and *actors' strategies* (Sect. 3.5)—*and* the local governance mechanisms that guide their interaction (Sect. 3.5) are at the core of the analysis in each article. A strong **theoretical connection** therefore exists between the four articles of this book (see Figs. 3.1, 3.2, and Fig. 5.1).

In each article, I analyzed housing as a human-made resource that is governed by the regulatory regime (SQ1) and the use strategies (SQ2) applied by different user actors (owners and non-owners). In each paper, my aim was to understand *how* and *why* power relations and games among these actors emerge to answer the main research question of *how* housing under scarce land conditions can be governed socially and sustainably.

**Article 1** starts with the analysis of the federal institutional regime in force to understand *how* the Swiss federal government defines policies for housing (re)development in dense urban environments (sub-question 1).



G. Debrunner, *The Business of Densification*,
https://doi.org/10.1007/978-3-031-49014-9_5

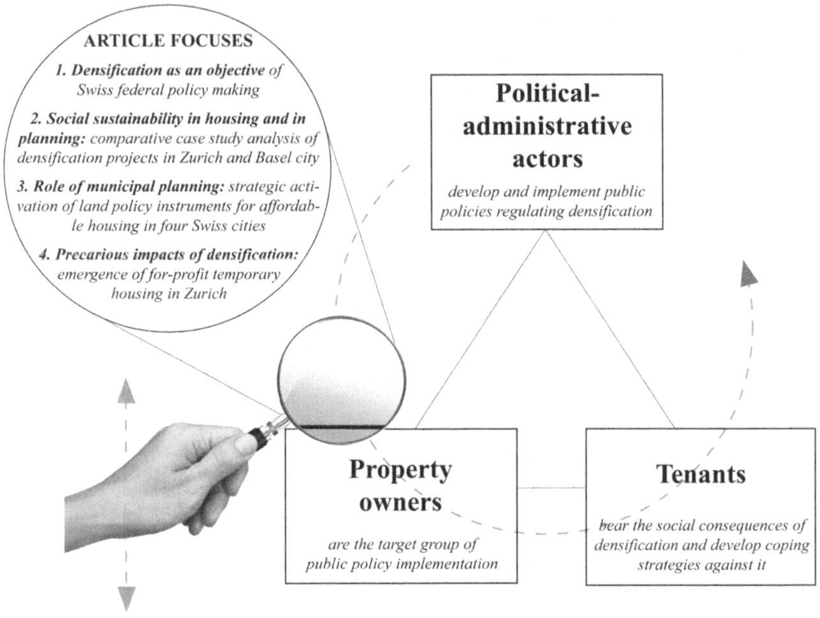


**Fig. 5.1** Book structure: actors' triangle and user appropriation strategies at different geographical scales

Our goal was to investigate how different actors use their power position to influence rule definition and formulation at the Swiss federal level. In particular, Andreas Hengstermann, Jean-David Gerber, and I investigated the implications of the Swiss federal policy context and the responsibility of national public actors (federal council, legislative parliament, lobby groups, etc.) in urban housing provision. We discussed the role of the federal state in attracting foreign investment for densification projects and in the protection of private property rights. In addition, we analyzed Swiss federal land use and housing policies, which are crucial for enabling access to and distribution of housing as a resource. We examined the institutional challenges and historical context of housing supply in Switzerland with a special focus on the social sustainability aspects of housing.

Thereby, the aim was to understand *how* the rights of private ownership are protected in the Swiss legal context and how this legal condition influences the use strategies employed by owner and non-owner actors. Finally, the analysis helped me to further elaborate how federal policy measures and instruments introduced trickle down and influence housing development also at the regional and municipal level. Conclusions were drawn regarding the extent to which the Swiss federal institutional regime exerts influence on the behaviors of specific actors, particularly on landowners. The article contributes to the critical examination of power relations in Swiss federal policymaking and in relation to the densification strategies applied by municipalities in an advanced capitalist liberal economy.

**Article 2** first discusses the concept of social sustainability in housing from different theoretical perspectives. It then examines *why* an emic approach towards social sustainability in housing is an added value for urban research. Through a comparative case study design, Arend Jonkman, Jean-David Gerber and I empirically investigated how social sustainability in housing is acknowledged by residents living in two large-scale densification areas in Zurich Brunau and Basel Schoren. We compared the social sustainability performance of these projects by using qualitative research methodology.

In this article, we brought together the different concepts—housing as a resource, institutions, and actors' strategies—and compared the local governance mechanisms for socially sustainable housing in two large-scale densification projects in the cities of Zurich and Basel. Besides the study of the strategies employed by public administrative actors, we also aimed to understand the resistance strategies of landowners *and* tenants in this paper. Moreover, we added to the understanding of how different local regulatory regimes contribute to conflicting outcomes in terms of social exclusion. Finally, we drew conclusions regarding the role and responsibility of public and private stakeholders in the local governance process for socially sustainable housing development in densifying urban areas.

**Article 3** further investigates *how* densification materializes at the municipal level. It focuses on public policies and policy instruments that guide densification of housing stocks in municipalities. More precisely, through comparative case study analysis Thomas Hartmann and I analyzed how land policy instruments can strategically be activated by municipal planning authorities to effectively promote socially sustainable housing in dense cities. While much literature so far has focused on the functioning of individual land policy instruments (e.g. land readjustment, growth boundaries, long-term leases), we expanded the existing body of literature in this field. We investigated how different instruments can strategically be combined and activated by local planning administrations.

Empirically, our focus was on the role played by municipal authorities and their applied land use and densification strategies in four Swiss municipalities (Zurich, Basel, Köniz, and Kloten). We concluded that municipal public authorities are indeed crucial for the decision-making process whether consolidation leads to social exclusion or not. They can alleviate rent level increases by strategically activating land policy instruments in favor of social inclusion, but they must therefore know *how* to densify. To prepare for future housing challenges, a stable "right-to-housing" for all does not necessarily require the mere introduction of new policy instruments, but the strategic activation of available instruments matters. By identifying the local governance mechanisms for social sustainability in housing, this article greatly helps municipal planners, practitioners, and policymakers to prepare for future housing challenges in dense urban environments.

**Article 4** shifts the focus from the federal to the local level and examines in a single case study how private property owners respond to the policy shift towards densification. In particular, Jean-David Gerber and I investigated *how* conflicting housing use interests under scarce urban land conditions enhance the landowners' interest to stay flexible in order to prevent building delays. We revealed that, when dealing with scarcity of land, landowners agree to enter a particular housing phenomenon. This model has evolved in the city of Zurich during the 2010s—the emergence of a profit-oriented temporary housing model that works under the rules of loaning law rather than rent. In this article, we analyzed an extreme example of precariousness in housing as a results of densification pressure. To cope with planning insecurity and building delays in the context of land scarcity, institutional owners (and the private real-estate industry working on their behalf) drove the emergence of this new business model.

By identifying the involved actors' objectives and strategies under the given legal frame- work in Zurich, this paper contributes to understanding *how* densification affects the interests of the lowest income segments. It shows that even in a city like Zurich—one of the richest globally—residents' social inclusion is not only a matter of affordable housing policy, but effective local governance in general. Finally, this article extends the existing literature on temporary use and housing by shedding light on how specific public *and* private actors are influenced by densification policies and *how* they are able to responding to it.

Table 5.1 provides an overview of the content investigated in each of the four research articles.

**Table 5.1** Overview of the four research articles composing this book

| | ***Article 1:*** *The Business of Densification—Distribution of Power, Wealth, and Inequality in Swiss Policy Making* | ***Article 2:*** *Planning for social sustainability: Mechanisms of social exclusion in densification projects in Swiss cities* | ***Article 3:*** *Strategic use of land policy instruments for affordable housing: Coping with social challenges in Swiss cities* | ***Article 4:*** *The Commodification of Temporary Housing in Zurich, Switzerland* |
|---|---|---|---|---|
| Status | **Published** in Town Planning Review, Vol. 91, No. 3, 259281 | **Published** in Housing Studies https://doi.org/10.1080/02673037.2022.2033174 | **Published** in Land Use Policy, Vol. 99, https://doi.org/10.1016/j.landusepol.2020.104993 | **Published** in Cities, https://doi.org/10.1016/j.cities.2020.102998 |
| Research questions | How are the mechanisms leading to a bypassing of tenants' rights in densification procedures in Switzerland to be explained? | How are tenants affected by densification from a social sustainability perspective?<br>How do the institutions in force contribute to explain the outcomes of urban densification projects in terms of social exclusion? | How do municipal planning authorities promote affordable housing in densifying cities? | What are the institutional arrangements making for-profit temporary housing possible?<br>Which rationales of the different actors involved in this system explain its expansion?<br>And what are the consequences on the different categories of actors? |
| Research design | Single case study analysis | Comparative case study analysis | Comparative case study analysis | Single case study analysis |

(continued)

**Table 5.1** (continued)

| | ***Article 1:*** *The Business of Densification—Distribution of Power, Wealth, and Inequality in Swiss Policy Making* | ***Article 2:*** *Planning for social sustainability: Mechanisms of social exclusion in densification projects in Swiss cities* | ***Article 3:*** *Strategic use of land policy instruments for affordable housing: Coping with social challenges in Swiss cities* | ***Article 4:*** *The Commodification of Temporary Housing in Zurich, Switzerland* |
|---|---|---|---|---|
| Unit of analysis | Swiss federal level | Swiss municipalities (Zurich Brunau, Basel Schoren) | Swiss municipalities (Zurich, Basel, Köniz and Kloten) | Swiss municipality (Zurich) |
| Autorship | Gabriela Debrunner Andreas Hengstermann<br>Jean-David Gerber | Gabriela Debrunner<br>Arend Jonkman<br>Jean-David Gerber | Gabriela Debrunner Thomas Hartmann | Gabriela Debrunner<br>Jean-David Gerber |

## 5.2 Article 1: The Business of Densification—Distribution of Power, Wealth, and Inequality in Swiss Policymaking

Gabriela Debrunner, Andreas Hengstermann & Jean-David Gerber
Journal: Town Planning Review, 2020, Vol. 91, No. 3, 259–281, https://doi.org/10.3828/tpr.2020.15
Impact Factor (2023): 1.8

Status: Published

**Abstract:** In Switzerland, the fight against uncontrolled urban sprawl and the protection of agricultural land have a long tradition. To reconcile these concerns, the Swiss voting majority agreed to introduce densification as a legally binding policy objective in 2013. Simultaneously, however, densification processes have started to threaten the housing situation of low- and middle-income tenants due to higher rents following redevelopment. In this article, we argue that the Swiss way of implementing densification is characterized by a systematic bypassing of tenants' needs for social sustainability in housing due to the current political priorities of the Swiss federal government. Using an institutionalist analysis approach and a qualitative case study methodology, we analyze the institutional mechanisms and the actors' rationale behind this emerging business of densification. Finally, we discuss the role of the nation state in the provision of the "right to housing" for all income segments and its consequences for the country's long-term sustainability performance.

*Keywords: densification, densification policy, housing, social sustainability, social justice*

**Research Highlights:**
*The paper…*

- *… analyzes how the revised Swiss Federal Planning Act (2013) and the introduced obligation to densify for municipalities has affected the housing situation in Switzerland.*
- *… summarizes the current state of densification policy making at the Swiss federal level.*

- *... applies an institutional analytical framework to identify intergovernmental elements of policy making in Swiss cantons and municipalities confronted with scarce land conditions.*
- *... questions the role of the Swiss federal council and the administrative offices working on their behalf in Swiss densification policy making.*
- *... examines how specific federal policies and instruments are implemented in favor of an "eco-business" of densification at the expense of its social side.*
- *... argues that a shift towards increased tenants' inclusion in federal policy-making and plan-ning is needed if public actors aim to support sustainable urban development effectively.*

### 5.2.1 Shift Towards Densification in Swiss Land Use Planning

In Switzerland, the fight against uncontrolled urban sprawl and the protection of agricultural land have a long tradition (Gennaio et al., 2009). To reconcile these concerns, densification has been introduced as legally binding policy objective in the revised Swiss Federal Spatial Planning Act (SPA)[1] in March 2013. Following the revised legislation, the 26 cantons and over 2000 municipalities must promote "inward settlement development, while ensuring an appropriate quality of housing" (Art. 1, para. 2, lit. $^{abis}$ SPA). Densification is thereby defined as a process leading to an increase in the number of housing units within existing municipal boundaries (Broitman & Koomen, 2015). It is hence assumed to play a decisive role in the fight against urban sprawl and the overuse of non-renewable resources (Swiss Federal Council, 2016a).

However, in the housing sector, densification arises with social challenges: considering the small size of the country and its constantly growing economy and steady population growth (+22% until 2045 [FSO, 2015]), densification of existing built-up areas has remained the main option to reconcile these concerns as the majority of greenfield and brownfield sites have been densified already (Nebel et al., 2017; Swiss Federal Council, 2017). Densification therefore increasingly materializes via "soft measures", e.g. in the form of total replacement constructions, modernization, subdivision, or conversion of existing buildings (Bibby et al., 2020). This soft way of implementing densification, in turn,

[1] Federal Act on Spatial Planning (SPA) of 22 June 1979 (CC 700).

has caused social exclusion and gentrification processes of residents as old housing stocks are being demolished and redeveloped with higher rents after densification (FOH, 2016b; FOSI & FOH, 2015). Low- and middle-income households face difficulties in finding adequate housing as newly modernized apartments are primarily affordable for households with higher incomes and non-profit housing suppliers have long waiting lists (FOH, 2016b). The situation is even more worrying considering that Switzerland has the lowest homeownership rate in Europe and is therefore regarded as a nation of tenants in a liberal housing market (Bourassa et al., 2010; Lawson, 2009). Housing provision traditionally lies in the responsibility of the profit-oriented private rental sector. 58% of Swiss households live in apartments of the for-profit rental sector and remain dependent on the private homeowner's decisions (FOH, 2017). As a consequence, resistance against densification projects has increased in recent years as tenants do not accept the social consequences caused by densification (Maissen, 2018; Swiss Federal Council, 2017).

In this article, we aim to demonstrate that the Swiss way of implementing densification policy is characterized by a systematic bypassing of tenants' needs for social sustainability in housing due to the current political priorities of the Swiss federal government. We explain the reasons and key actors' rationales behind this difference of treatment between tenants' and homeowners' rights. Finally, our objective is to discuss the role of the national state in the provision of the "right to housing" for all income segments and its consequences for the state's sustainability performance as a whole.

To answer these research objectives, we rely on an institutionalist approach. This allows us to analyze densification as a process that emerges within a tight web of diverse and contradictory rights and regulations. Its performance and implementation depend on veto rights controlled by powerful actors as well as intertwined private and public interests. We hence proceed in three steps to analyze these mechanisms between densification policy implementation and housing development at the federal level: (1) Through a broad screening of the institutional regime in force, we examine the policy measures taken in federal legislation promoting the tenants' housing needs under densification. (2) We reconstruct the rationale behind the policy measures applied through an analysis of actors' interests, objectives, and policy strategies. (3) We evaluate how housing is being addressed in federal densification policy and draw conclusions on potential repercussions on spatial planning, social justice, and sustainable urban development.

### 5.2.2 *Densification at the Interplay Between Public Policy and Property Rights*

For portraying the housing challenges that arise through densification adequately, the article applies an institutionalist approach (Healey, 2007a; Jessop, 2001; Mandelbaum, 1985). In general, this approach assumes that the actual spatial development (e.g. situation on the housing market) is to be seen as results of the institutional setting and vice versa actors pursuit their interests by activating specific formal and informal rules from this institutional regime (Dembski & Salet, 2010; Hall & Taylor, 1996; Hartmann & Gerber, 2018). Specifically, in this article, the intertwined relationship between densification and housing will be explained by an analysis of the Institutional Resource Regime (IRR) in which institutions—defined as a set of rules and regulations—their effectiveness and formulation process also play a central role (Gerber et al., 2009). Without denying the importance of informal rules, focusing on formal rules in particular, the IRR builds on the assumption that—besides public policies—institutions in general (Hardin, 1968, 1991; North, 1994; Williamson, 2000) and property rights in particular (Bromley, 1992; Demsetz, 1967; Ostrom, 1990, 2009) are central for understanding (un)sustainability of resources. It hence postulates a causal relationship between institutions, actors' behavior, and condition of resource and enables to explain the social consequences of densification as a result of behavioral patterns stemming from incentives of different policy fields—especially in regard to the attenuation of private property rights (Gerber et al., 2009; Knoepfel et al., 2007).

Following the IRR approach, two main categories of formalized rules guide the implementation of densification processes—public policies and property rights—that operate according to a different logic and rely on opposing legitimisations.

- *Public policies* aim at solving a public problem recognized as such by democratic processes. Through public policies, the state receives the power to regulate the actions of those actors who are thought to be at the source of the problem, in the name of the public interest.

Public policies are regularly revised, not only because the problem they are targeting constantly evolves, but also because changing political majorities propose alternative solutions to the problem (Knoepfel et al., 2011).

- *Property rights* protect individuals from the state (Marx, 1868; Rousseau, 1762). As such, they defend private interests against the (potentially absolutist) action of the state (Locke, 1689). Property rights are grounded in the Civil Code. They are extremely stable over time because their definition hardly changes.

*Property rights* and *public policies* interact in a dialectical relationship. Property rights make the private appropriation of goods and services provided by resources possible, as long as public policies do not restrict exclusive appropriation in the name of the public interest(s). By doing so, public policies have a redistributive effect (Knoepfel, 1986).

In the last three decades, densification as a public policy objective has found its way into legally binding regulations (Williams et al., 2000). Densification policies are defined as a set of rules, with the common interest to solve the politically defined problem of urban sprawl through the process of increasing density (number of housing units) within the existing boundaries of built-up areas (Broitman & Koomen, 2015; Knoepfel et al., 2007). Besides "command and control" policies directly addressing the extent, range, or type of uses, such as land use planning, housing, environmental and monument protection laws, also more indirect public interventions based on incentives or information play a role in densification (Hood, 1983).

However, decision-making procedures in densification policy implementation are complex due to intricate small-scale ownership structures, veto rights controlled by landowners who can block (or slow down) implementation processes, and intertwined private and public interests (Dempsey, 2010; Holman et al., 2015). Public authorities may influence the property owner's decision whether to raise rents after redevelopment or not (e.g. through targeted policy intervention in housing, tax, or energy law) (Slaev, 2016), but it is ultimately on the landowner to decide on the level of profitability to be targeted (Buitelaar & Needham, 2007). Without heavy state intervention such as expropriation, new planning regulations (e.g., new zoning) only get implemented when titleholders agree to undertake new developments, sell their land, or transfer their development rights (Davy, 2005, 2012). Therefore, public policies with a spatial impact often conflict with the landowners' freedom (Blomquist, 2012; Gerber et al., 2017; Slaev, 2016).

In the formulation process of densification policy objectives, public and private actors strategically use their policy resources (such as knowledge, money, or personnel) to enforce their interests and to achieve their political goals. For instance, actors use their widespread network to achieve consensus in the introduction of a new densification rule. Simultaneously, actors organize and develop political strategies within the given institutional setting to regulate their access to a resource (e.g. affordable housing). To reconstruct the rationale behind the densification policy measures applied, we examine the strategies of public and private stakeholders involved in the densification policy process at the federal level, their contribution to the formulation of institutional rules and policy objectives, as well as their ability to address housing needs in federal policy intervention (Knoepfel et al., 2007).

### *5.2.3 Housing from a Social Sustainability Perspective*

In this article, housing is regarded as a human-constructed resource (Kébir, 2010). According to the Universal Declaration of Human Rights (Art. 25), having a home is a basic human need and an essential good. Its conditions of access, supply, management, distribution, and ownership structure need to be properly thought through to avoid conflicts between competing uses (e.g. affordable housing and lofts for couples with double income), which has consequences for its sustainability (Bathelt & Glückler, 2005). As a key element of the built environment, housing under densification also becomes particularly relevant for sustainable development as a whole (Chiu, 2003).

In recent years, a number of government and academic reports have indicated that the different dimensions of sustainability in housing development have not been equally prioritized by policymakers (Chiu, 2004; IFHP, 2019; Vallance et al., 2011). The International Federation of Housing, for instance, only recently stated that "social sustainability is the most neglected element of the three because it is far more difficult to quantify, contextualize and develop than economic growth or environmental impact" (IFHP, 2019). It hence is an ambiguous and fuzzy concept that comes with a number of ethical, political, and methodological challenges (Weingaertner & Moberg, 2014; Woodcraft, 2012).

Therefore, our objective is not to expand the list of social sustainability definitions (see for a discussion Bramley et al., 2009; Chiu, 2003; Davidson, 2010; Dempsey et al., 2011; Littig & Griessler, 2005; McKenzie, 2004; Polese & Stren, 2000; Vallance et al., 2011). Rather we argue that housing affordability is one of the key elements of urban social sustainability. Particularly, in a context of densification, social displacement of low- and middle-income residents—as central defining trait and primary danger of gentrification—increasingly emerges due to higher rents after densification processes (Lees et al., 2008). Through the affordability of rents, in contrast, residents get a chance to stay in their neighborhood, which is why other social sustainability criteria of housing development such as accessibility, residential stability, tenure security, local identity, or community cohesion also get preserved (Bramley et al., 2009; Chiu, 2003; Vallance et al., 2011). Therefore, housing affordability is a good indicator of social sustainability of housing as a whole. By definition, housing affordability relates to the cost of housing relative to household income and other legitimate expenses (Mulliner et al., 2013: 275). Evidence from Switzerland additionally shows that housing affordability plays a central contribution to the people's social acceptance of densification and hence to sustainable development as a whole (COSD & CSO Zurich, 2014).

In the end, however, the judgment, evaluation, and reflection on whether social sustainability in densification processes is given or not will be made along broader social (in)justice principles (see also Jehling & Hartmann, this issue). Rawls (1972: 303), for example, states that a socially just city is designed in a way that equally and inclusively distributes the rights, chances, and opportunities among all people of a society. His understanding of social justice refers to the need to improve the life prospects of the least advantaged by fostering affordable and secure housing conditions, their involvement in formal decision-making, and their access to attractive open and public spaces to reduce social inequalities. In the conclusion section of this paper, we reflect on this principle of social justice related to the densification policy measures applied.

### 5.2.4 *Case Study of Densification Policies in Switzerland*

Switzerland makes an interesting case study to analyze the relationship between densification policy implementation and housing (re)development, as the challenge of coordinating the two has become critical in the country in recent years. Specifically, issues of housing affordability and gentrification have intensified in almost every Swiss city in recent years due to population growth and national densification objectives (FOH, 2016a, 2016b; Rérat, 2012; Wehrmüller, 2014). At the same time, yield-oriented investments attracted by the country's economic stability and wealth, reinforced the attractiveness of Swiss real-estate markets. Because of low-interest rates, housing has become the main target of capital investment, especially for pension funds (Theurillat et al., 2014). Consequently, in Swiss cities, the housing situation is characterized by an overheated housing market (vacancy rates below 1% and rising rents) (Balmer & Gerber, 2017). Within the rental market, 28.9% of households suffer from excessive housing costs in relation to income (FOSI & FOH, 2015).

Hence single case study analysis allows to assess and to explain the diverse mechanisms behind densification policy and housing (re)development in detail (Yin, 2018). The federal state plays a crucial role in this matter in the sense that it signals how to deal with this issue also for cantons and municipalities. Switzerland is organized on three executive levels (municipalities, cantons, and the confederation) and characterized by a form of "cooperative federalism". Legislation in favor of densification goals and/or social sustainability objectives of housing is introduced by the federal state and is to be implemented by cantons and municipalities (Linder, 1994).

We employed qualitative methods to understand the relationship between densification as a core objective of public policy and its consequences for social sustainability in the housing sector. In the first step, about 40 policy documents were analyzed. Legally binding documents (e.g. acts, ordinances, changes in legislations, and vote results) were considered as well as policy documents without a legally binding nature (e.g. strategy papers, government reports, parliamentary debates, and position papers) disclosing the composition of political arenas, leitmotifs, and parliamentary debates. In the second step, the political negotiations behind federal legislation were revealed by conducting nine semi-structured expert interviews with public and private representatives at the federal level from three federal offices, the Swiss homeowners' association, the Swiss tenants' association, the Swiss association of the building industry,[2] as well as the Swiss association of institutional investors. Two members of the national council were additionally interviewed because of their detailed understanding and knowledge of the topic.

### 5.2.5 *Densification Policy Implementation and Social Sustainability in The Swiss Housing Sector*

In this section, we analyze the institutional regime (stressing both public policies and property rights) regulating densification processes in Switzerland. Densification is addressed in several public law areas. We start with the Swiss housing policy. Then, we emphasize aspects of planning policy because of their significant impact on housing (re)development. Three additional sectoral policies that influence sub-aspects of housing densification are also explored. Finally, the role of private law will be addressed. Lastly, we reconstruct the rationale behind federal policy measures through an analysis of actors' interests, objectives, and policy strategies.

*Screening of Institutional Rules Addressing Social Sustainability of Housing Under Densification in Swiss Legislation*

Housing Policy

Unaffordable housing prices are a central problem targeted by housing policies. Swiss housing policy is anchored in two constitutional articles. They prevent abuses in tenancy matters (Art. 109 CSC)[3] and regulate the supply of affordable housing (Art. 108), particularly for disadvantaged groups (e.g., elderly, disabled, and low-income households). In 2003, based on the constitutional mandate "to meet the housing needs" (Art. 108), a new Federal Housing Support Act (FHSA) was introduced to support the construction and renewal of affordable housing, as well as the activities of non-profit housing organizations (Art. 2 FHSA)[4] (Balmer & Gerber, 2017). However, direct loans granted by the federal government were suspended due to the 2003 "federal budget relief program" (Swiss Federal Council, 2014).

[2] In German: Verband «bauenschweiz»—Schweizerische Dachorganisation der Bauwirtschaft.

[3] Federal Constitution of the Swiss Confederation (SC) of 18 April 1999 (CC 101).

[4] Federal Housing Support Act (FHSA) of 21 March 2003 (CC 842).

Since then, only indirect support mechanisms in favor of non-profit housing organizations (e.g., housing cooperatives) and state guarantees on bonds issued by non-profit housing developers have been implemented. These loans or advantageous mortgages can only be obtained if the non-profit housing organization commits to the cost-rent principle and belongs to an umbrella association that promotes non-profit housing (Lawson, 2009). In total, the share of non-profit housing developers in Switzerland (public and cooperative) reaches 6% of the total housing stock (FOH, 2017). Benefit payments for tenants (demand-side housing subsidies) exist in Switzerland, but only in the context of social assistance and state supplementary benefits to old age and disability insurance. In the absence of additional public subsidies, since construction costs cannot be reduced directly, newly built non-profit housing is only affordable to households with a medium income, not to the poorest segment of the population (Balmer & Gerber, 2017). Due to long waiting lists, the time to get access to a cooperative housing unit usually takes several months or years (Burri, 2015).

Planning Policy

Through land use planning policy, the state aims to promote the sustainable use of land in its economic, social, and ecologic dimensions. In Switzerland, as a reaction to the immense construction activity in the decades following the Second World War, the need for coordinated spatial development became predominant. In 1969, Swiss citizens therefore approved a constitutional amendment adding spatial planning to the list of official state powers. The overarching planning objective and legitimacy is "to ensure the appropriate and economic use of the land and its properly ordered settlement" (Art. 75 CSC). Thereby, planning gained the competence to limit private construction activity. Zoning plans became mandatory throughout the country, separating building from non-building areas. In recent years, political and professional debates questioned whether this approach of outward limitation was sufficiently effective. In March 2013, the Swiss voting majority therefore agreed to a reinforcement of the Federal Spatial Planning Act (SPA) introducing, inter alia, densification as a legally binding policy objective. Following the revised legislation, cantons and municipalities must arrange "settlements according to the needs of their inhabitants and their expansion must be limited" (Art. 3, para. 3 SPA).

Energy Policy

Through its energy policy, the state coordinates the use of energy in settlements and controls forthcoming environmental and socio-economic consequences. Switzerland's "sustainable use of energy" is anchored in two constitutional articles (Art. 89, 91 CSC). To meet this objective, Switzerland revised its Federal $CO_2$ Act (2011)[5] and Energy Act (2016)[6] promoting the shift towards sustainable energy transition. In total, by 2020, domestic greenhouse gas emissions are to be reduced by a total of 20% as compared to 1990 (Art. 3 $CO_2$ Act). To achieve this goal the country must renovate its existing building stock as a whole to successfully reduce its energy depletion by 2050. This can either be done through total replacement constructions or energetic renovations (Vonmont, 2016).

This quantitative policy goal directly affects densification processes in the built environment as existing buildings have to be redeveloped to fulfill the emission limit. Therefore, the $CO_2$ Act was supplemented by two policy programs, namely, the "building programme" (in force since 2010) and the "Federal Energy Strategy 2050" (in force since 2018). The building program is financed via the $CO_2$-levy on fuels (Art. 34 $CO_2$ Act) and serves as a public subsidy pool promoting energy efficient building renovations. Private homeowners can submit funding proposals to municipalities and, in return, are directly funded by the federal state (e.g. for the insulation of windows, facades or photovoltaic systems up to 30% of their total investment). Today, Switzerland's granting of subsidies is explicitly based on criteria improving energy efficiency. The market situation and the different needs of investors or residents are not included in the analysis (FOH, 2016a).

Tax Policy

Equal treatment in the provision of taxes to the community is the core objective of the state's tax policy. The Swiss federal tax system, characterized by extensive finance and revenue sharing between the confederation,

[5] Federal Act on the Reduction of $CO_2$ Emissions ($CO_2$ Act) of 23 December 2011 (CC 641.71).

[6] Federal Act on Energy (FEA) of 30 September 2016 (CC 730).

the cantons, and the municipalities, is anchored in several constitutional articles (Art. 3, Art. 126–135 CSC). Based on the Federal Act on Direct Federal Tax[7], private investments carried out in buildings to improve energy efficiency have been considered equivalent to maintenance costs and are therefore seen as tax-deductible for a long time (FOH, 2016a). With the implementation of the new Energy Act, by the year 2020, it will even be possible to deduct taxes for renovations, redevelopments, and demolitions up to three years after completion. The aim here is to encourage total renovation instead of partial renovation, which until 2018 has been more attractive in fiscal terms (Federal Tax Administration, 2017). Moreover, investors can additionally benefit from so called "deadweight loss effects", meaning that they declare tax deductions for renovations although their investment would be profitable without public support. According to a federal interdepartmental study of 2009, these deadweight loss effects amount to 70–80% of the total amount of tax reliefs for energy-saving measures in Switzerland (FDF et al., 2009).

Heritage Protection Policy

The heritage protection policy aims to preserve buildings, sites, or landscapes with a specific value for society. These values can be of historic, architectonic, aesthetic, political, ideologic, or economic nature. In Switzerland, "the protection and preservation of historic landscapes and buildings" is anchored in two constitutional articles (Art. 10 & 78 CSC). It is stipulated in the Federal Act on the Protection of Nature and Cultural Heritage (NCHA)[8] that three federal inventories regulate the fulfilling of this task: the federal inventory of landscapes and natural monuments, the inventory of Swiss cultural heritage sites, and the inventory of historic pathways and transport routes. In recent years, however, the scarcity of land for development, population growth, and increasing mobility have started to threaten the existence of preserved and historic sites in Switzerland. Heritage conservation objectives are increasingly under pressure, both for economic and environmental reasons: protective inventories hinder the full use of economic potentials and energy objectives are more and more aimed at being achieved regardless of the protection status of the area (Swiss Federal Council, 2018). According

[7] Federal Act on Direct Federal Tax of 14 December 1990 (CC 642.11).

[8] Federal Act on the Protection of Nature Cultural Heritage (NCHA) of 1 July 1966 (CC 451).

to a recent study of the Federal Office for Culture (2018), however, heritage protection massively influences the social dimension of urban development. Through the protection of monuments, social livelihoods, the cohesion of neighborhoods and the local identity tend to remain preserved. This, in turn, can also have an influence on housing prices as older buildings are generally more affordable to low-income segments than modernized units (FOC, 2018).

Property Rights

*Property owners' rights* are well protected in Switzerland. The Swiss Constitution protects the "right to own" as a fundamental right (Art. 22ter CSC) that can only be restricted if (1) a legal basis and an overweighing public interest exist, (2) the measure is proportional, and (3) a full compensation is paid (Art. 5, Art. 36 para. 1–3 & Art. 26 para. CSC). In practice, the weight of public interest is interpreted narrowly by courts so that property restrictions are limited and expropriations are rare in international comparison (Alterman, 2010).

*Swiss tenants' rights* are protected by articles for tenancy matters in the Swiss Constitution (Art. 109 CSC) as well as the Federal Obligations Code[16] of 1911 (Art. 253–274 OC). Swiss tenancy law is acknowledged to be weak in comparison to neighboring states (e.g. Germany) (GFOBRP, 2016). For instance, private landlords are allowed to terminate an open-ended rent contract without any legal restriction at any time. Swiss tenancy law also allows property owners to pass energy-saving investments on to the tenants for up to 50–70% of the total costs. Although it is stipulated in the implementing legislation (Art. 14 OC) that landlords who have received public subsidies for renovations must deduct them from the new rent, in practice, energy-related renovations are considered to be equivalent to maintenance costs and legitimize rent increases (FDF et al., 2009). In addition, besides the investment costs, homeowners can also pass energy costs (e.g., for heating) and the $CO_2$-levy to the tenants (FDF et al., 2009). Nevertheless, within the existing housing stock, tenancy law does not allow for continuous rent increases. Landlords have to align their existing rents with the current interest rate that corresponds to the average mortgage rates. Interestingly, since the newly introduced densification rule in planning law in 2013 (Art. 1, para. 2, lit. abis SPA), the number of demolitions and replacement buildings has increased significantly as this procedure has remained the only chance for property owners to bring the rent to a higher level within the existing housing stock.

After all, Swiss tenancy regulation works on the basis that tenants have to claim their rights in the cantonal tenancy court. Only if they defend their rights in court, can they make themselves visible to landlords and public authorities. In practice, however, tenants often do not use this option because they need to remain on good terms with their landlord to secure further housing offers or they do not have the resources to do so.

To sum up, the analysis of public policies reveals that so far the social sustainability dimension of housing development has been neglected in Swiss densification policy implementation. While planning and energy policy exclusively pay attention to the ecological dimension of densification, e.g. by introducing new legislation that enshrines energy objectives, legal amendments addressing socially sustainable densification were not made—neither in housing, tax, nor heritage policy. This observation can also be confirmed when looking at private law regulations: property rights remain strongly protected in Switzerland and no changes in tenancy law have been made although the conditions for tenants in the housing segment changed significantly in recent years. Especially since new densification and energy policy objectives have been introduced in 2013. In other words, the legal regime in force does not adequately accredit the housing needs of the people mostly affected by densification—the tenants.

Table 5.2 summarizes the federal policy instruments in force addressing social sustainability of housing in a context of densification. Additionally, Table 5.2 incorporates political arguments and policy initiatives in the national council aiming to introduce instruments in favor of such social sustainability measures. In the subsequent section, we discuss the reasons and actors' rationales for the rejection of these proposals.

*The Rationale Behind Swiss Densification Policy Measures*

In the following section, we reconstruct the rationale behind the Swiss densification policy measures applied. The involved actors' motivations for the introduction of the policy instruments described above will be explained in detail. Moreover, the actors' objectives when rejecting proposed initiatives in the national council in favor of social sustainability of housing densification will be made intelligible.

The Private For-Profit Rental Industry

**Table 5.2** Formal federal institutions and policy instruments impacting social sustainability of housing under densification. The data analysis is based on an in-depth study of policy papers, legislations, and parliamentary debates (Fluri, 2017; FOC, 2018; FOE & FOH, 2015; FOE, 2018; FOF et al., 2009; FOH, 2012, 2014a, 2014b, 2014c, 2016a, 2016b, 2017; FOSD, 2016, 2018; FOSI & FOH, 2015; FTA, 2017; Hardegger, 2017; Swiss Federal Council, 2014, 2016a, 2016b, 2017, 2018; von Graffenried, 2014)

| *Formal institutions* | *Federal policy instruments in force impacting housing development under densification* | *Federal policy instruments in favour of socially sustainable housing development that are not in force but have been politically debated* |
|---|---|---|
| **Housing policy**<br>**Swiss Constitution (Art. 2, 41, 108, 109)**: Social objectives, housing construction, tenancy matters<br>**Legislations:** Federal Housing Support Act (2003) | • Supply-side subsidies ("fond de roulement" and direct loans for non-profit housing developers) | • Increase of the direct loans provided in the Federal Housing Support Act in favor of low-cost housing<br>• Introduction of supply-side subsidies for all non- and for-profit developers<br>• Introduction of a fixed minimum share of non-profit housing in the Swiss Constitution |
| **Planning policy**<br>**Swiss Constitution (Art. 75)**: Spatial planning<br>**Legislations**: Revised Spatial Planning Act (SPA) (2012, Art. 1, 3, 8a, 15, 15a, 38a) | • Mandatory conformity to cantonal plans and local zoning regulations<br>• Densification as legally binding planning objective for all cantons and municipalities<br>• Sustainable housing development (ensuring adequate quality of housing)<br>• Capturing planning-related added value<br>• Building defined as subject to approval | • Introduction of low-cost housing in SPA<br>• Introduction of municipal pre-emption rights in favor of low-cost and nonprofit housing construction<br>• Introduction of a minimum densification quota in renovation and new development projects<br>• Introduction of occupancy rules to reduce the consumption of living space per person |

(continued)

**Table 5.2** (continued)

| *Formal institutions* | *Federal policy instruments in force impacting housing development under densification* | *Federal policy instruments in favour of socially sustainable housing development that are not in force but have been politically debated* |
|---|---|---|
| **Energy policy**<br>**Swiss Constitution (Art. 74, 89)**: Environment protection and energy policy<br>**Legislations**: $CO_2$ Act (2011)<br>Revision of Energy Act (2018) | • Direct supply-side subsidies for private homeowners when doing energetic renovations via the federal building program (Art. 34, CO2 Act) | • Introduction of mandatory "building energy certificates" in all cantons<br>• Introduction of the mandatory disclosure of "consumption-based heating and hot water costs" in all cantons<br>• Public subsidy support only if certain rental rates are achieved (e.g., no dismissal after redevelopment) |
| **Tax policy**<br>**Swiss Constitution (Art. 3, 126–135)**: Federal tax system<br>**Legislations**: Federal Act on Direct Federal Tax (1990) | • Tax relief on home ownership and private property (assets)<br>• Tax relief for self-occupied property ("own rental value")<br>• Will be introduced in 2020: Tax relief for renovations, redevelopments and demolitions of buildings up to three years after having finished the task | • Obligation for private homeowners to pass tax benefits for energetic renovation to the tenants<br>• Cancelling deductions from taxable income when doing energy renovations<br>• Trading with emission certificates between the cantons |

| *Formal institutions* | *Federal policy instruments in force impacting housing development under densification* | *Federal policy instruments in favour of socially sustainable housing development that are not in force but have been politically debated* |
|---|---|---|
| **Heritage protection policy**<br>**Swiss Constitution (Art. 10, 78)**: Preservation of Landscape and the built environment<br>**Legislations**: Federal Act on the Protection of Nature and Cultural Heritage (1966) | • Federal Inventory of landscapes and natural monuments<br>• Federal Inventory of Swiss heritage sites<br>• Federal Inventory of historic pathways and transport routes | • Facilitation of protective measures in the Federal law (would lead to less heritage protection in settlements) |
| **Property rights**<br>**Swiss Constitution (Art. 26, 36, 109):** Property guarantee, Tenancy matters<br>**Legislations:** Civil Code (1907, Art. 641), Obligations Code (1911) | • Property rights, leases, mortgages<br>• Tenancy law: Obligation to terminate an openended rent contract with a minimum of 3 months before dismissal (but: at any time and with no reason possible; no mandatory right for rent extension or compensation for the tenants)<br>• Obligation for homeowners to pass received public subsidies for renovation to the tenants;<br>• But: the costs of the whole renovation can at the same time be passed to the tenants up to 50–70% of the total investment | • Reduction of the transfer rate from owners to tenants in case of renovations<br>• Introduction of a mandatory disclosure of the former rent on behalf of the tenants<br>• Introduction of a dismissal protection of tenants in case of rising rents after redevelopment |

Since 2014, the private for-profit rental industry (including for-profit investors as well as private planning, building, and architecture professionals) has increasingly started to appreciate densification as a profitable investment market. Reasons are that they (1) are publicly subsidized for carrying out renovations, (2) do not face any obligations to pass on received subsidies to the tenants, and (3) simultaneously can take advantage of higher rents after redevelopment. Institutional investors, for example, have intensified their work with professional planning, building, and architectural teams to better exploit the new use potentials and to increase their land rent in the existing housing stock. Through densification, they can enlarge the rentable floor space on a parcel and expand their investment opportunities at central locations. The better they understand how to use and upgrade the existing housing stock through densification, the higher the prices that will be paid.

> Extensions to existing buildings, additional floors, or total replacement constructions with double volume. On the same property, owners can suddenly realise a lot more things. In the end, it's all about the land price and about making money. (Vice-director Swiss Association of the Building Industry, August 2018)[9]

Therefore, densification measures have provided a clear legal and economic incentive for the private for-profit rental industry to increase rents.

Interestingly, private small-scale owners expanding and renting out their owner-occupied properties have recognized their economic advantages too. Before 2013, this share of private homeowners strongly criticized the shift towards densification as they feared the loss of property value, privacy, and autonomy. This skepticism has waned in recent years, mainly due to the mentioned institutional rules and amendments applied in energy, tenancy, and tax law (see §6.5.2). As a result of these legislative changes, property owners of the for-profit rental industry (both institutional and owner-occupied property) assess the economic benefits of densification higher than the risks and costs, which is why they both increasingly agree to densify their parcels. In fact, densification under the new legal framework and in the absence of further restrictions (e.g. in tenancy law) fulfills a specific function on real-estate markets in the sense

[9] All quotes have been translated to English from German by the authors.

that it increases planning and economic security through the increased opportunity to invest into real estate. In an environment where land is scarce and competition to use this land is high, property owners reinvest their assets into stable and safe investment markets. In the end, because of the weak protection of tenants, owners enjoy maximal planning flexibility and decision-making power in densification projects.

This economic interest also explains why political efforts to change the institutional rules in force in favor of protective and price-regulating objectives (e.g. in housing or planning policy) have been rejected by the private for-profit rental industry. From a private landowner's perspective, further regulation would make the planning process too complicated, time-consuming, and costly. This would reduce their investment security and their willingness to densify their parcels.

> You cannot have both without restrictions: densification and heritage protection. Thereby, densification will be more difficult to implement because for ¾ of the projects a special legal approval would be needed. Therefore, at the national level, we demand for a weakening of the monument pro-tection regulations. For instance, that buildings from the 1960s no longer need to be preserved. (Head of the legal and planning department of the Swiss Homeowners' Association, August 2018)

In other words, the private for-profit rental industry tends to support a way of implementing densification that guarantees economic growth and income at the expense of preserving social values (e.g. cultural heritage).

In addition, in recent years, this industry has increasingly realized how to use their legal power position when negotiating with public authorities about densification projects. Due to the strong protection of their property title, they hold the power of disposal and grant the use rights. Therefore, public authorities are increasingly dependent on the private owners' agreement when aiming to implement densification objectives. Urban planning regulations such as zoning, capturing of planning-related added value, or tax relief suddenly become negotiable for private individuals and do not represent binding rules any longer.

> Yes, we [public with private actors] negotiate with each other. […] However, one would first need to think about whether this [the legal obligation to densify] still is a binding rule if the best solution becomes negotiable for both sides. (Vice-director Swiss association of the Building Industry, August 2018)

As a matter of fact, this power position in the negotiation process results in densification measures primarily being implemented along market-oriented principles in Switzerland.

The Swiss Federal Government

After the Swiss voting majority agreed to introduce densification as a legally binding planning objective in 2013, Swiss federal government has become responsible for the fulfilling of this task. Since 2014, the responsible federal offices (mainly the office for spatial planning, but also for housing and energy) have started to convince the private for-profit rental industry to support the implementation of densification as a policy objective through dedicated policy measures. Concretely this means that although several new densification objectives in planning and energy law have been introduced during the 2010s, federal administrative authorities have not obligated the private for-profit rental sector to return any of the received public subsidies to the tenants. For example, private landowners do not have to fulfill a certain quality related to socio-economic living standards (e.g. house prices, social mixing, residential stability) or to pass on tax savings or direct subsidies to the tenants. Social issues such as housing affordability, social mixing or security of tenancy have not gained political attention even though the housing situation for low-income has worsened in recent years. Swiss tenancy law, in addition, has not been revised even though owner-related and tenant-relevant legislative changes have been made in energy and planning law. Furthermore, the private for-profit rental industry has neither been forced to protect tenants from dismissal or rent increase after redevelopment, nor have political efforts in favor of monument conservation policies received support of the political majority (e.g. to preserve architectonic and social qualities within the built environment).

The rejections of such initiatives were justified by the fact that federal authorities feared a decrease in the property owners' interest to renovate and to carry out energy-saving renovations. It was also argued that private property owners would not have been able to use the existing economic potentials on their parcels the same way anymore. The legal support of residential stability and heritage protection in the built environment would have led to increasing financial expenses, distorted market conditions, and the loss of the country's position as an attractive real-estate market for international investment. As a result, social sustainability objectives such as housing affordability, residential stability, and neighborhood cohesion have not gained the support of the national council due to the strong liberal interests of the private for-profit rental sector.

> For densification you get a political majority anyway. But for social sustainability you get none. […] Because densification brings money and investment opportunities. […] As long as one can make economic profit, densification will be supported by the majority of people. There is a coalition between ecologic Switzerland and the liberal interests, and if both can be combined, a political majority prevails. (Member of National Council and former director of the Swiss Tenants' Association, August 2018)

The tenants' needs related to social sustainability have remained bypassed under the new rules of the game (densification). This prioritization in policymaking is also connected to the poor representation and lobbying position of the tenants' interests in the national council.

> The stepchild in the whole debate is the social dimension. From a political point of view also badly organised. People suffering from poverty do not have a political lobby at the federal level […]. Bringing the losers together to defend their political interests is difficult. (Director of the Swiss Federal Office for Housing, August 2018)

In fact, the tenants' and property owners' rights are unequally represented in the national council. This imbalance of power distribution due to the tenants' lack of access to formal decision-making (missing lobbying position) reinforces the trade-off in favor of the economic and ecologic dimensions of sustainability in densification processes at the expense of its social side. As a result, densification policy measures leading to higher income, use potentials, and increased land value for private property owners (through economic and ecologic densification policy measures) remain prioritized.

### *5.2.6 Towards a Business of Densification*

In this article, our aim was to demonstrate that the Swiss way to implement densification is characterized by a systematic bypassing of tenants' need for social sustainability in housing due to present-day political priorities of the Swiss federal government. Even though the Swiss federal council publicly commits to socially equitable sustainable development

(Art. 2, 41, 74 CSC and Art. 1 SPA), in densification measures, it deliberately assists in bypassing the social dimension of sustainability in favor of eco-economic development objectives. This strategy is connected to the fear that densification might not be effective due to the power position of the landowners involved in the for-profit rental industry. For instance, federal administrative authorities have recognized a decline in the private for-profit rental industry's willingness to invest into (re)development projects, if they would further limit and regulate densification at the federal level. Therefore, they follow a strategy that is profitable enough for the profit-oriented rental sector to get densification implemented. Simultaneously, the landowners' profit margin in densification projects has increased significantly in recent years due to the economic and legal security, stability, and predictability provided by the national state. It appears that the powerful lobby of property owners was able to promote specific policy measures (e.g. in the domain of environmental sustainability) to reinforce their own benefit and financial returns. Consequently, in comparison to the 1990s and 2000s, the private for-profit rental sector has started to acknowledge densification as a new, safe, and profitable investment market, which is why they represent the winners of the emerging "business of densification". The tenants, in contrast, remain excluded from the compromise made by powerful actors and embody the losers of the new rules of the game (densification). In particular, low-income tenants such as elderly people, families, and migrants cannot afford higher rents after densification measures and are increasingly being excluded from the housing market. Because only a small share of Swiss tenants gets publicly subsidized through social aid, or benefits of public or cooperative housing supply, a growing percentage of the Swiss population suffers from inadequate housing supply in relation to income.

The emergence of the business of densification is related to a general shift towards the commodification of housing in many Western societies: the value of housing is more and more considered by its financial value at the expense of its use value (Harloe, 1995; Rolnik, 2013). This supports the constant erosion of stability and security in housing. An increasing number of households lack access to adequate housing on the regular rental market. Regarding the future increase of densification projects, these vulnerable groups will be caught in a vicious situation and depend on housing solutions that will inevitably lead to the erosion of their social rights, stability, and protection in housing (Brenner et al., 2012; Harvey, 2012; Marcuse, 2012). The Swiss case additionally shows that the tenants' lack of access to formal decision-making at the federal level (missing lobbying power) reinforces social inequalities in Swiss housing development: tenants have to live with a double burden as they increasingly pay for environmental costs even though they are not the only producers and suffer the related quality of life burdens such as rising rents, the instability of tenancy, and the potential loss of neighborhood cohesion. In the meantime, the federal state withdraws from its responsibility to cover the housing needs for all income segments and passes the duty to the cantons, the municipalities and, ultimately, to the individual households. As a consequence, municipalities are increasingly responsible for mitigating social risks, challenges, and problems (Heeg, 2013).

Overall, we argue that the Swiss federal government tends to underestimate the potential consequences and challenges for future sustainable development as a constitutional objective resulting from neglected social sustainability—especially in housing as a key component of the built environment. Sustainability only makes sense if its social dimension is taken seriously. Otherwise the sustainability objective as a whole is missed. The way densification is currently being implemented in Switzerland leads to an exacerbation of landlord–tenants relations, but this is seen as the necessary price to be paid for improving energy efficiency and environmental development objectives. The social living quality for all segments of the population, however, can only be maintained if future densification projects are compatible with the interests of culturally and socially diverse groups, and at the same time encourage social integration and more equitable distribution of decision-making power. Otherwise, short-term profitability objectives will take the upper hand and rents will increase in such a way that densification will be rejected by the majority of the population due to the lack of affordability and social acceptance. We identify an evident risk that the implementation of sustainable development objectives through densification initiatives might slow down—or even come to a standstill—because the residents' housing and social needs are not seriously taken into account. We argue that new legal ways need to be found to better consider the housing needs of those who are the most affected by densification but who are often not in a position to be heard.

## 5.3 Article 2: Planning for Social Sustainability: Mechanisms of Social Exclusion in Densification Through Large-Scale Redevelopment Projects in Swiss Cities

Gabriela Debrunner, Arend Jonkman & Jean-David Gerber
Journal: Housing Studies, https://doi.org/10.1080/02673037.2022.2033174
Impact Factor (2019): 2.255

Status: Published

**Abstract:** In many countries of the global North, there has been renewed interest over the last 30 years in urban densification as part of wider efforts to combat urban sprawl and the loss of natural habitats. In daily practice, however, densification is a contested process because of its redistributive effects. Next to potential environmental advantages, it produces both benefits (e.g. increased housing options, business opportunities) and losses (e.g. rising noise, rents) for different individuals and households. The redistributive effects are an expression of conflicts between environmental, economic, and social dimensions. In this article, we show the latter is heavily impacted: if densification projects are not designed to the needs and capabilities of people who are actually supposed to benefit from it—the residents—low-income groups are at risk of discrimination, displacement, and exclusion. A scenario which is highly unsustainable. By using a neoinstitutional approach and comparative case study methodology conducted in Zurich and Basel, Switzerland, we analyze the institutional rules and the involved actors' strategies (local authorities, investors, tenants, NGOs) when dealing with urban densification projects. We explain the mechanisms at play leading to the loss of a city's social qualities when competing with short-term economic interests of investors and local authorities.

*Keywords: densification; new institutionalism; housing; social sustainability; gentrification*

**Research Highlights:**
*The paper…*

- *...introduces a neoinstitutional analytical framework and explains how this helps to systematically analyze housing challenges in a context of densification.*
- *...performs a comparative case study approach to compare the land policy strategies of two Swiss cities – Zurich and Basel – both confronted with scarce urban land conditions, affordable housing shortages, and increasing densification pressure.*
- *...introduces criteria for the evaluation of densification projects from an emic housing sustainability perspective.*
- *...compares the sustainability performance of two large-scale densification projects in the cities of Basel and Zurich.*
- *...critically analyzes and reflects on the role of the city government and the landowners when undertaking densification projects.*

### *5.3.1 Implications of Densification on Social Sustainability in Housing*

In many cities of the global North, tensions between densification as a policy goal and its social implications on housing affordability, residential stability, or community cohesion have intensified in recent years (UN Habitat, 2016). In Switzerland, for example, municipalities have become obliged to promote densification within existing municipal frontiers to protect agricultural land and to prevent urban sprawl since May 2014 (Art. 1 SPA). Simultaneously, a growing number of tenants living in rental housing stocks is confronted with the situation of being evicted and displaced at short notice as they cannot afford the rents after densification and simultaneous modernization anymore (FOH, 2019: 4).

We identify two lines of research related to the social implications of urban densification in the rental segment: first, a broad body of literature reflecting on the pros and cons of densification, both as a process and policy objective (e.g. Holman et al., 2015; Touati-Morel, 2015). And second, scientific work discussing the role of social sustainability in urban regeneration in general (e.g. Ancell & Thompson-Fawcett, 2008; Arthurson, 2001; Bramley et al., 2009; Bramley & Morgan, 2003; Burton, 2000, 2003; Chiu, 2004; Vallance et al., 2011; Marcuse, 2016). However, a critical analysis which focuses on the socio-political dimensions of densification and its effects on tenants from a social sustainability perspective is largely missing (Pérez, 2020). As we will argue in the following sections, such understanding is crucial so that densification projects are actually designed in a way that takes into account the needs and capabilities of those affected—the residents—and involve them into decision-making. Otherwise, especially low-income groups will not be able to participate in urban development without being vulnerable to discrimination and likely to suffer displacement and exclusion from their communities (Jenks et al., 1996: 84; Scally & Tighe, 2015).

In this article, our goals are twofold: *first*, we aim to explain how tenants are affected by densification from a social sustainability perspective to understand the social consequences at the households level. Thereby, we are in line with emic research approaches that argue that the principal source of evidence concerning the sustainability of cities should be people themselves, particularly those living in the areas in question (Bramley et al., 2009; Jenks et al., 1996; Vallance et al., 2011; Zukin, 2009). *Second*, our goal is to detect the reasons for tenants' social exclusion in densification projects from a neoinstitutional perspective. More precisely, we analyze the local regulatory framework and the strategies of the actors involved (local authorities, investors, tenants, NGOs) to understand the mechanisms at play that potentially hamper a socially-sensitive implementation of densification. Specifically, we ask: *1) How are the impacts of urban densification on tenants to be analyzed from a sustainability perspective? And 2) How do the institutions in force contribute to explain the outcomes of urban densification projects in terms of social exclusion?*

These questions require the use of qualitative case study methodology (Yin, 2018). We conducted a comparative analysis of two Swiss cities—Zurich and Basel. Both cities are confronted with increasing densification pressure and tenants exclusion due to rising rents after redevelopment. By analyzing the institutional rules and decision-making behavior of the actors involved in two large-scale densification areas, we explain the reasons for possible trade-offs between economic, environmental, and social goals of densification. We show that preserving the cities' social qualities is in acute danger when competing with short-term economic interests of investors and local authorities. Finally, we discuss our results in regard to arising repercussions for Swiss urban policymaking and planning.

### 5.3.2 *Planning for Social Sustainability in a Dense City*

Densification is defined as a process leading to an increase in the number of households living within existing city boundaries (Broitman & Koomen, 2015: 32). In many cities, the process has been introduced as a legally binding policy objective during the 1990s to effectively steer efficient use of natural resources and "smart growth". In daily practice, densification within municipal boundaries may materialize in different forms: for example, via infill on empty sites, conversion of buildings used for other functions, or complete demolishment and reconstruction including more housing units of existing housing stocks (Touati-Morel, 2015). In municipalities in which free inner-city parcels are missing and there is a lack of space to relocate other functions, the latter option gains in relevance.

*Evaluating the Impacts of Urban Densification on Tenants from a Social Sustainability Perspective*

"Sustainability" depends on the interaction of economic changes with social, cultural, and ecological transformations. If one of the dimensions is not adequately secured, the development cannot be considered sustainable (Barbier, 1987: 103).

Vallance et al. (2011: 344) argue that the residents' interpretation of the local environment is central for measuring sustainable development. They state that if the social preconditions to support densification are not given—resulting in rising housing prices—an urban development scenario is highly unsustainable (Jenks et al., 1996: 84). Densification, in other words, needs to respect the "places" and "spaces" in which tenants live and are socially embedded in to preserve the city's long-term social stability and capital (Lefebvre, 1991). "[C]ities cannot be considered sustainable if they are not acceptable to people as places in which to live, work, and interact or if their communities are unstable and dysfunctional" (Bramley et al., 2009: 2125). Thus, for densification to be truly sustainable, it has to esteem tenants basic needs and the specific social relations, values, customs, and structures of the place they live in (Chiu, 2004: 66). This residents-oriented sustainability approach (Chiu, 2004; Townroe, 1996) is linked to actual urban development practices at the local level rather than to broad initiatives, policy agendas, or policy objectives. It acknowledges that social sustainability is indeed a community level concern, but depends on the extent to which individuals can contribute to it (Elsinga et al., 2020). The approach is guided by the conviction that the principal source of evidence concerning the social sustainability of cities should be people themselves, particularly those living in the areas in question (Bramley et al., 2009; Zukin, 2009).

Although we agree that it is impossible to provide comprehensive universal sustainability standards given the socio-cultural and geographical diversities of human settlements (Chiu, 2004: 75), we argue that a more detailed understanding of the social side of densification is crucial in reconciling the often competing demands of the society–environment–economy tripartite (Vallance et al., 2011: 342). Although it is more than thirty years since the Brundtland report's release and extensive academic literature has been published on the concept of sustainable development, its social dimension has only received little attention in policy, academia, and practice (Manzi, 2010; Murphy, 2012). However, social sustainability research is needed to add to existing understanding and perceptions of sustainable development as all three dimensions of sustainability are interlinked. The diminishing of one affects that of the others (Chiu, 2003; Khan, 1995; Mitlin & Satterthwaite, 1996; Williams et al., 2000).

While each of the indicators of social sustainability (Table 3.2 of this book) may be regarded as conceptually distinct it is clear however that there exist various reinforcing relationships between them (Chiu, 2004: 65). The indicators introduced were obtained by synthesizing selected academic and policy literature with the ambition of highlighting key aspects of interest for social sustainability in relation to densification and urban housing development from a tenants perspective. They were supplemented by our own experience working with residents, local authorities, housing suppliers, and community organizations in Switzerland and the Netherlands.

*(see Table 3.2 of this book)*

The *affordability* of housing is the key dimension with regard to the social sustainability of housing for households (Jonkman, 2020; Yung & Lee, 2012). A household's ability to meet the cost of housing is the core limiting factor as to whether they can access adequate housing or not (Ancell & Thompson-Fawcett, 2008: 432). Moreover, the *availability* and *quality* of housing are also crucial to assess (Mulliner et al., 2013). For example, in many cases, residents are able to afford housing but they still remain excluded from the housing market, e.g. through the limited availability of housing or discrimination. Housing availability refers to the situation whether apartments in the required price range are also available at the time designated. This issue becomes particularly relevant when many rental contracts have been terminated simultaneously in the same area (IFHP, 2019). In addition, the quality of housing is of central importance when issues of overcrowding, inadequacy, and poor design impact people's lives (Ancell & Thompson-Fawcett, 2008). It describes whether residents live in housing conditions that fail to meet physical standards of decency or to be situated at unsafe or inaccessible locations (Stone, 2006). Moreover, it expresses an array of attributes (e.g. access to services and facilities), in addition to purely economic factors that can influence a household's perception of affordability (Mulliner et al., 2013). Finally, *community cohesion* is used as an indicator to describe the level of residents' social attachment to the local community. Particularly, a stable community is regarded as a necessary capability of a community to sustain itself (Chiu, 2004; Dempsey et al., 2011). *Citizenship* describes the residents inclusion to local decision-making which provides information on whether the tenants' needs and perspectives are integrated also on a formal level (Bramley & Power, 2009; Fainstein, 2010).

*Explaining Social Exclusion from a Neoinstitutional Perspective*
Supporting tenants' social inclusion in densification projects has largely been assumed to be the responsibility of the public sector, more specifically of local authorities as they guide, structure or even determine the use of urban space (Holman et al., 2015). As Healey (2007a) highlights, however, the social impacts of densification are to be seen as the results of a complex process of governance which is to be understood as the interplay between the local regulatory framework and the decision- making behavior of the actors involved. Following the neoinstitutional perspective, human actions take place within a tight web of institutional rules which structure humans' expectations about what others will do (Hall & Taylor, 1996: 956). Institutions are defined as shared social values stipulated in formal laws and ordinances (e.g. in planning law) which guide social interaction and practices (Dembski & Salet, 2010: 612). Within this institutional setting, actors (e.g. local authorities, landowners) develop strategies to defend their own interests in order to meet a particular policy goal (e.g. densification) (Gerber et al., 2018: 11).

Besides public officials, other groups such as lobby parties, landowners, developers, and residents do also play a crucial role in the decision-making process. These actors influence whether gentrification processes after modernization of housing stocks emerge or not. Each of them can support social issues through the strategic activation of specific formal rules. Landowners, for instance, are most often in a position of power due to the protection guaranteed by private property rights. On private plots, public action only gets implemented when titleholders agree to undertake a new development, sell their land, or transfer their development rights (Gerber et al., 2018). Consequently, in many cases, the landowner is free to define the profit margin to be targeted on the parcel and can set the rents according to market-prices. Such commodification strategies (Aalbers, 2017; Marcuse, 2016), however, may hamper tenants' social inclusion and result in the promotion of housing based on its financial value rather than its use value (Rolnik, 2013).

To sum up, the socially-sensitive implementation of densification is the result of a socio-political negotiation process which is shaped by the local regulatory framework stipulated in formal rules (e.g. legislations, codes, ordinances) and the strategic behavior of the actors involved (Nicol & Knoepfel, 2008). Codominant use interests between residents, investors, and local authorities and their strategic formulation and activation of specific formal rules result in benefits for some (e.g. increased housing options, business opportunities) and losses for others (e.g. displacement, insecure tenure, community disruption) (Brenner et al., 2012; Marcuse, 2016).

### 5.3.3 *Comparative Case Study Analysis in Zurich Brunau and Basel Schorenweg*

To analyze a contemporary phenomenon—the challenging implementation of densification objectives in terms of social sustainability—within its real-world context, we conducted in-depth qualitative case studies (Yin, 2018).

*Case Selection*

In Swiss cities, the tensions between densification objectives and tenants' interests have become predominant in recent years, especially since the revision of Federal Spatial Planning Act (SPA) in 2013 obliging over 2000 municipalities to densify within city boundaries. An increasing number of people suffer from social displacement after modernization as a consequence of densification (FOH, 2019: 4). As the country is regarded as a nation of tenants with the lowest homeownership rate in Europe

(Lawson, 2009), a growing number of inhabitants living in the private rental market is at risk to be evicted at short notice due to decisions taken by the landowner (Rérat, 2012). In Switzerland, the municipality is the actor responsible to coordinate densification. Local planning authorities grant the building permits to private landowners. Building applications need to align with the Local Zoning Plan.

We selected two comparative cases—the Swiss municipalities of Zurich and Basel—to analyze two different governance approaches towards socially-sensitive densification. In both cities, the pressure on housing development under scarce land conditions has risen in recent years: in Zurich, for example, where 28% of the population lives in apartments of the private rental market (Table 5.3), the number of densification projects of private investors has quadrupled since 2006 (from 9 to 36 projects in total). The absolute number of dismissals due to densification measures in the city's private rental sector has doubled within the period of 2006 to 2017 (Statistics City of Zurich, 2017). As a result of increasing housing prices after modernization and densification, moving to cheaper suburban areas remains the only option for lower-income groups in both municipalities (Balmer & Gerber, 2017).

*Project Selection*

To evaluate the social sustainability in urban densification from a tenants' perspective (research question 1), we further selected two large-scale densification areas within Zurich and Basel city area. In this project-based approach, we investigated social sustainability "from the ground up, as it actually exists in local places, and as a set of evolving practices" (Krueger & Agyeman, 2005: 416). Specifically, the densification projects—Zurich Brunaupark and Basel Schorenweg—were selected as they are both owned by the same institutional investor (Credit Suisse bank [CS]). This player is representative for many urban residential areas in Switzerland since the share owned by institutional investors such as CS makes 63% of the total housing property in Swiss cities (FOH, 2017: 14). CS' projects in Zurich and Basel were both ongoing at the time of investigation (between March and November 2019) which is why the actors involved (local authorities, investors, tenants, NGOs) could be directly confronted with their decisions and actions taken.

**Table 5.3** Housing market characteristics in Zurich and Basel-City (FOSD, 2017: 25; Statistical Offices Zurich and Basel-City, 2019, 2020)

| | *Population in absolute numbers (2019)* | *Estimated population growth & demographic change* | *Share of buildable lots within city boundaries (2017)* | *Owner-occupied (including condominium) (%)* | *Private rental housing (%)* | *Social rental housing (incl. public housing) (%)* |
|---|---|---|---|---|---|---|
| City of Zurich | 428′700 (around 1 Mio. including suburban areas) | +21% by 2030, significant increase of 10–19 and 40–49 years old | 4% (for housing purposes only) | 47.2 | 28 | 24.9 |
| City of Basel | 178′445 (around 800′000 including suburban areas) | +10% by 2040, significant increase of children (10–19 years) and 40–49 | 4% (for housing purposes only) | 49.7 | 36.1 | 13.5 |

Zurich Brunaupark—a settlement built in the 1980/90s—compounds of four buildings with 239 apartments and approximately 400 residents. The investor plans to densify the area through demolishment and total reconstruction of the existing buildings in 2023. The new settlement will count an additional 258 apartments, 497 in total (Schoop et al., 2020: 18). Many households (47%) consist of multiple adults without children. 42% of the households have been living in the project for over 15 years. Basel Schorenweg—built in 1961—counts 196 apartments with around 300 residents in total. CS plans to densify the two existing buildings via total internal reconstruction with smaller housing units in 2021 (Laur, 2019: 21). In March 2019, the approximately 1085 tenants in total were informed of the termination of their rental contract by CS. A high share of the residents can be classified as old-aged and/or as single-households (Table 5.4).

*Methods*

The data of our study was conducted through qualitative methods. We proceeded in two steps: *first*, we analyzed how tenants living in the areas in question (Brunaupark and Schorenweg) are affected by densification from a social sustainability perspective. We conducted a household survey with 412 households living in the settlements in total to gain a broad understanding of their perspectives. The survey incorporated the social sustainability indicators presented in the previous section and included both open and multiple-choice questions. The open questions were used to gain a rich understanding of the households' perspective on how tenants are affected by densification plans. The multiple-choice questions were used to further underline household positions, but the analysis remains qualitative in nature. We opted for a self-completion postal and digital survey method (with one reminder) and managed to achieve a respectable 25% response rate (101 responses in total). In designing the questionnaire, we considered the existing body of literature as well as a number of national surveys covering similar topics which helped us to identify whether and how questions have shown to work.

In a *second step*, we analyzed the reasons for possible trade-offs between the tenants', landowners', and local authorities' interests focusing on institutional rules and actors decision-making behavior (neoinstitutionalist perspective). In this step, we started with a broad screening of local policy documents to analyze the interface between urban densification and social

**Table 5.4** Socio-economic profile and household types of tenants in Zurich Brunaupark & Basel Schorenweg

| | *Zurich Brunaupark (%)* | *Basel Schorenweg (%)* |
|---|---|---|
| *Socio-economic-profile and household types* | | |
| Single person, under 35 | 3 | 7 |
| Single person, 35 to 65 | 15 | 10 |
| Single person, 65 or older | 8 | 43 |
| Two or more person household (no children), all under 35 | 7 | – |
| Two or more person household (no children), not all under 35 or over 65 | 25 | 12 |
| Two or more person household (no children), all 65 or older | 15 | 12 |
| Couple with children, youngest child 6 or younger | 12 | 7 |
| Couple with children, youngest child 7 or older | 8 | 7 |
| Single parent, youngest child 6 or younger | – | 2 |
| Single parent, youngest child under 7 or older | 5 | – |
| *Years of residence in the settlement* | | |
| <1 year | 7 | 5 |
| 1–4 years | 5 | 24 |
| 4–10 years | 25 | 24 |
| 10–15 years | 20 | 17 |
| >15 years | 42 | 31 |

sustainability. We included government reports, legislation, and parliamentary debates primarily being published within the last decade. We also incorporated newspaper articles, project documents, and “grey literature” to understand the actors’ strategies and objectives behind specific formal rules activated. Finally, we employed ten semi-structured expert interviews with representatives from five local public authority departments, three local tenants associations, and two CS portfolio managers. All experts were chosen due to their detailed understanding and knowledge of the topic. The data was evaluated using qualitative analysis methods.

### 5.3.4 *The Tensions Between Densification and Social Exclusion in Swiss Urban Policymaking*

In the following section, *first*, we show how tenants are affected by densification from a social sustainability perspective in Zurich Brunaupark and Basel Schorenweg. *Second*, to understand the reasons for specific decisions and policy measures taken in each institutional setting, we analyze what strategies local authorities, property owners, and tenants develop to defend their interests in urban densification.

*Impacts of Densification on Tenants from a Social Sustainability Perspective*

Community Cohesion

In Zurich Brunaupark and Basel Schorenweg, the majority of the residents has lived in the settlement for over 15 years (Table 5.4). Many of them state that they feel strongly socially embedded in the neighborhood as they have spent their everyday life with families, children, and friends together and share a lot of memories. In particular, families with children as well as old-aged fear to lose social support and contacts through dismissal. They perceive a common sense of home, local identity, and embeddedness and are not willing to leave.

> We live in a small village here. People know each other. Everyone helps each other, talks together, meets in a coffee shop or in the local grocery store. We have a good social life and connectivity.[...] We live together very peacefully and quietly. (Tenant Zurich Brunaupark, 73 years, June 2019)

Housing Affordability

According to Credit Suisse' marketing department, the rents after densification and modernization in Brunaupark will increase by 60%. For example, a 3.5 rooms apartment (75 $m^2$) which today costs 1700 CHF per month (gross rent) will be offered for 2720 CHF. In Basel Schorenweg, the rents for the new apartments will rise by +50%. Here, a 3.5 rooms dwelling which costs today 1200 CHF per month (gross rent) will be offered for around 1800 CHF. In both projects, the bank legitimizes the rent increase with the argument that the dwellings are centrally located and substantial modernization resulting in higher living quality for the residents will be obtained (Credit Suisse Zurich & Basel, 2020).

As a consequence, tenants in Brunaupark and Schorenweg state that they will not be able to afford a new apartment in the modernized housing project anymore. Especially low-income and old-aged which have lived in their dwellings for many years indicate that they will have to move to cheaper areas outside city boundaries.

> I will lose my center of life. I will not be able to find an affordable apartment at such a central location anymore. (Tenant Basel Schorenweg, 55 years, June 2019)

Housing Availability and Accessibility

In Zurich, evicted tenants which are in need to find something at a low cost within the city (e.g. due to their workplace) rely on the support of non-profit housing associations. Otherwise, rents on the regular housing market are too expensive for them. In the city of Zurich, however, waiting lists for social housing units are long. People sometimes have to wait for several months to years to get access to an available low-cost apartment (Martel, 2020). Even if they are old-aged or in a precarious living situation, available apartments in the social housing sector are not offered to socially evicted tenants immediately or with prior access criteria. Consequently, for the majority of tenants living in Zurich Brunaupark, moving to a cheaper suburban area remains the only option to find housing.

> We will not find such an affordable flat in the city center anymore. All cooperative housing associations have long waiting lists and for some it is even not possible to apply anymore. [...] We are a community which is now being disrupted. (Tenant Zurich Brunaupark, 42 years, June 2019)

Even tenants with higher incomes who would be able to afford higher rents are not allowed to stay in Brunaupark. They do not receive prior access to the new dwellings even though they have lived in the settlement for many years since CS has decided to start with new residents from anew regardless of the former residents' family situation, age, gender, income, or workplace (Interviewee 39, CS portfolio manager Zurich, July 8, 2019).

Similarly as in Zurich, in Basel it has become difficult for evicted renters to find something adequate within the city center (Statistics Basel-City, 2019: 13). To find affordable housing at short notice, tenants also rely on the support of Basel's social housing associations. These social organizations, however, have long waiting lists too and do not prioritize old-aged, families or socially dismissed households (Martel, 2020). Similarly as in Zurich, even tenants who would afford the new rents in the densified settlement will not be able to stay in Basel Schorenweg. Neither they receive an alternative apartment which they could move to during reconstruction nor will they get prior access to a new dwelling. Therefore, regardless of being high- or low-income, moving to retirement homes (which have long waiting lists too) or to cheaper suburban areas remains the only option for tenants living in Basel Schorenweg (Beck & Schulthess, 2019).

Housing Quality in and Around the Building

From a socio-ecological point of view, tenants living in Zurich Brunaupark do not recognize a need for energetic modernization. The buildings have been internally renovated eight years ago. Specifically, in 2012, new sanitary facilities (kitchens, bathrooms) as well as new floors were installed (Schoop et al., 2020: 18). Therefore, tenants perceive the physical condition of their dwellings and the surrounding neighborhood as of high quality. In particular, they are satisfied with the size, the location, and the services within and around Brunaupark.

> I totally cannot understand why these buildings which are in a very good physical shape will be demolished. Especially in Zurich municipality which aims to reach the goals of a green and sustainable city. (Tenant Zurich Brunaupark, 78 years, June 2019)

Similarly as in Zurich Brunaupark, residents in Schorenweg do not understand why urban regeneration of their apartments is needed at all. In 2002, the buildings have been fully internally renovated. The modernization included the installations of new bathrooms, kitchens, floors, and windows insulation to improve energy efficiency. In 2010, in addition, the roof and gutters were renewed and in 2015 the eight elevators have been fully refurbished (Laur, 2019). Hence, residents living in Schorenweg perceive the physical condition of their apartments of high construction and housing quality. They also appreciate the access to green and open spaces as well as to services in the surrounding neighborhood. Thus, overall, "no construction measures are effectively needed" (Interviewee 37, Head of Local Tenants Association Basel, February 13, 2019).

> The building is in a very good condition. Densification will lead to modernization which only rich people can afford. (Tenant Basel Schorenweg, 65 years, June 2019)

Citizenship and Decision-Making

Finally, tenants of Zurich Brunaupark and Basel Schorenweg do not feel being adequately involved in the local decision-making process. Neither have they been informed about the upcoming dismissal in advance, nor have they been involved in the negotiation process between the city government and the investor from the beginning. For example, until contract termination, they have not known about the upcoming densification procedure and rent increase at all. The communication explicitly took place between CS and the city authorities.

> The property owner only communicates with the city council. The developer communicates with the tenants but only little and in a non-sympathic way. (Tenant Zurich Brunaupark, June 2019)

Therefore, tenants in both settlements do not feel adequately supported by the city council (executive) and local public administration. They feel to be left alone in finding a new apartment and in coping with their current living situation.

> We were surprised when we received the contract termination. Our government just observes and does not intervene. (Tenant of Basel Schorenweg, June 2019)

Overall, the results of the surveys show that the indicators of social sustainability are not met in both densification areas. The tenants are neither able to afford the apartments after densification, nor do they manage to maintain their social networks and acquaintances. They are forced to leave their dwellings even many of them face difficulties to find alternative housing options in the city. The densification procedures strongly disrupt their social stability and cohesion in the neighborhood. Results also show that tenants perspectives have not been formally addressed either. The decision whether, how, and for the benefit of whom densification was actually needed was explicitly made between the investor and the city council. This shows that urban planning in the age of densification does not or only insufficiently take the interests of the residents into account, even though they are very vulnerable to spatial changes.

*Institutional Rules and Actors' Strategies in Zurich and Basel-City*
In the following section, we analyze the institutional mechanisms leading to the situation presented in the previous section. To understand the reasons for the social outcomes identified, we analyze the institutional rules and the involved actors decision-making behavior. For each city, first, we emphasize aspects of planning and energy policy because of their significant impact on housing (re-) development. Second, objectives of housing and social welfare policy are also explored. Finally, we address the role of private law (property rights, tenancy matters) (Fig. 5.2).

Zurich City Government
On November 1st 2018, the revision of Zurich's Local Zoning Plan[18] came into force. Based on the revised legislation, the city council initiated planning measures such as the introduction of densification zones[19] to effectively promote population growth through internal settlement development and the efficient use of energy (Zurich City Council, 2013: 5). In Zurich Brunaupark, for example, the revised zoning plan has led to a situation in which the investor became allowed to double the number of apartments on the same parcel (by +258 additional apartments to 496 in total). The city council legitimized the introduction of these planning measures with the political aims to support future population growth through densification and to meet green policy objectives. In particular, the redevelopment of existing housing stocks is needed since free inner-city brownfield areas are missing in Zurich and new construction on greenfield has become restricted since the introduction of the revised Federal SPA (City of Zurich, 2016, 2019b).

> In future terms, and nowadays already, population growth is only possible through demolition and reconstruction of existing housing stocks in the city of Zurich. The municipality has to ensure that sufficient housing units will be provided. (Interviewee 40, City of Zurich, Urban Development Department, July 31, 2019)

Simultaneously, the fulfilling of social policy objectives such as the provision of affordable, stable, and secure housing is guaranteed by the

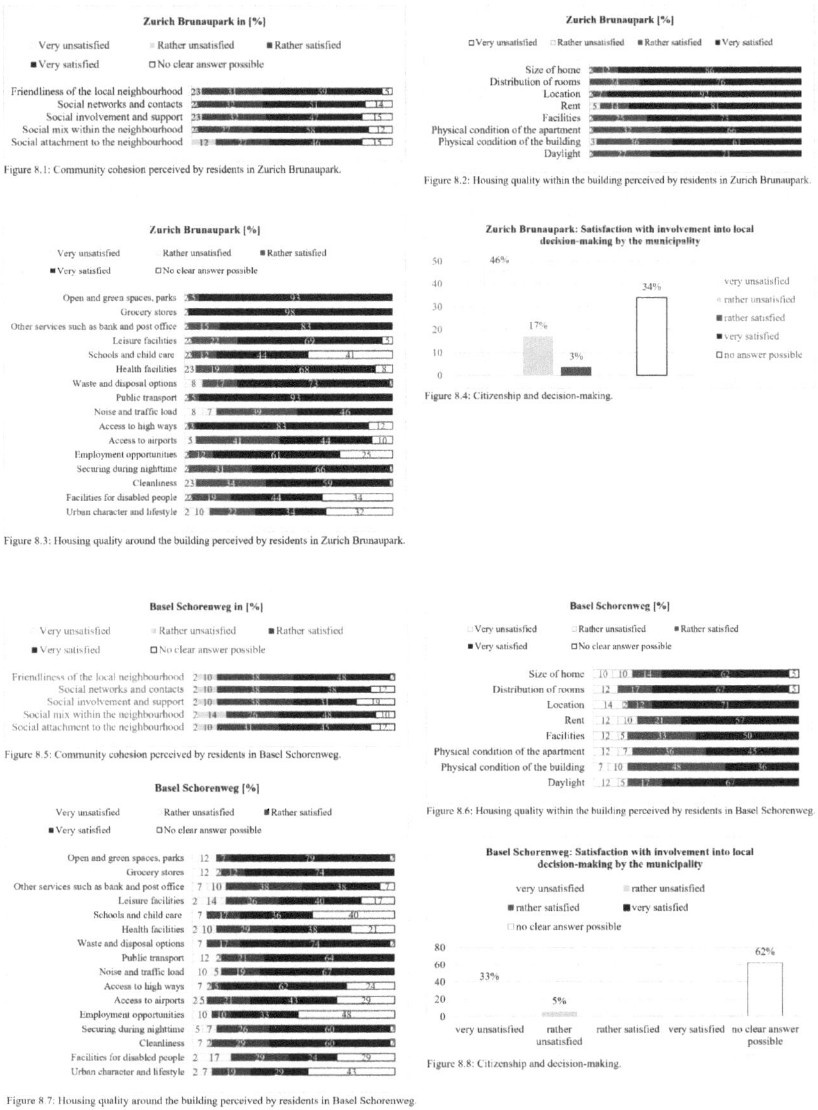


**Fig. 5.2** Overview of the results of the household survey conducted in Zurich Brunaupark and Basel Schorenweg. Results show how tenants are affected by densification from a social sustainability perspective

"Housing Article" in the Constitution of the Municipality of Zurich.[10] According to the Municipal Constitution (Art. 2, Para. 4), by 2050, the city council must ensure that a third of the total housing stock will comply in non-profit housing property and align with cost-rent principles to counteract social exclusion processes.

> There is a process of social exclusion going on in Zurich. If housing property has been renovated, demolished, and brought to the market again, the price for the same apartment with a higher standard has doubled. [...] We have a constant struggle of gentrification in the city. (Interviewee 35, City of Zurich, Housing Department, May 3, 2019)

To achieve this goal, during the last decade, the city government has followed an interventionist housing policy strategy to promote affordable housing e.g. by purchasing private land for public housing or by providing long-term building leases on public land and supply-side subsidies to non-profit cooperatives (Interviewee 40, City of Zurich, Urban Development Department, July 31, 2019). Overall, with these housing and planning policy measures, Zurich's local government aimed to constantly increase the share of social housing property within city boundaries (Zurich City Council, 2017: 4ff.).

> The municipality of Zurich follows an active housing policy approach. We are strongly linked to investors, private homeowners, and housing cooperatives and aim for frequent exchange. (Interviewee 45, Head of Planning Department, October 24, 2019)

As our analysis reveals, however, in daily practice, the promotion of "social sustainability" criteria (e.g. residential stability) does not only rely on local housing and planning policy. In Zurich, the building permit for each private housing project is approved and controlled by the municipal planning department. In case the project is of certain importance (e.g. due to its location) and size, the authority may receive recommendations of the Local Building Committee [LBC] which is part of the planning department (Art. 53 MC). The committee consists of external experts (mainly architects) as well as employees of the local administration. Its role is to advise the city council and the local planning department in

[10] Zurich Municipal Constitution of 24 November 2013 (MC 101.100).

questions of urban planning, design, and architecture. In Zurich Brunaupark, for instance, the committee advised the authorities to approve a total area reconstruction rather than partial redevelopment to ensure an improved and uniformed architectural quality of the settlement (Zurich City Council, 2019a). The fulfilling of social (and distributive) tasks e.g. in relation to tenants inclusion, community cohesion, or housing affordability was not part of their project evaluation.

> In Zurich, this [the collaboration with the local building committee] is called 'cooperative planning'. However, in Brunaupark, they only evaluated the projects based on design standards. Social parameters were not included at all. (Interviewee 31, Head of Local Tenants Association Zurich, April 18, 2019)

Moreover, the implementation of social objectives does not rely only on the local government's own prerogatives. This is because in Switzerland, in general, the rights of private homeowners are strongly protected by law in international comparison (Property Rights Alliance, 2019). At the municipal level, this means that the right to own property is protected as a fundamental right which can only be restricted if an overweighing public interest exists (Art. 22ter CSC[11]). As a holder of property rights, CS not only has the right to control and to make decisions about the housing stock in Brunaupark. It also has the right to obtain at least a portion of the financial benefits produced by the housing stock. In the rental sector, Swiss courts interpret the weight of public interest narrowly so that property restrictions or expropriations are rare in international comparison (Alterman, 2010). The rights of tenants (Art. 253–274 OC), in contrast, are regarded as weakly protected by law in comparison to neighboring states such as Austria or Germany (GFOBRP, 2016). For example, landowners are allowed to terminate an open-ended rent contract within three months without any specific reason. So regardless of the tenants' strength of social integration, age, or years of residency in the neighborhood. In Brunaupark, CS does not need to introduce rent levels for the new housing construction and is allowed to set the new rents according to market prices. They also do not need to follow legal restrictions for the dismissal of old-aged or economically weak households.

[11] Swiss Civil Code of 10 December 1907 (CC 210).

> Swiss tenancy law has one fundamental problem: in case of renewal, there is a divided housing market in Switzerland between tenants which have lived in their apartments for many years and the ones moving into new housing units. The former will not be able to move out of their current apartments since they will not be able to afford the rents offered on the regular housing market anymore. (Interviewee 31, Head of Local Tenants Association Zurich, April 18, 2019)

Finally, these private law restrictions have led to a situation in which the local planning department is only able to guide housing development to a limited degree. To a large extent, it relies on the responsibility of CS to decide on the profit margin and social goals to be targeted on the private parcel.

> Normally, the property owner has already decided whether they demolish the housing stock or not. The only thing we can do is to advise them in case the location is sensitive. We cannot do more than this. (Interviewee 45, Head of Planning Department, October 24, 2019)

Based on these legal conditions, on June 12, 2019, Zurich's executive city council decided not to approve the objections submitted by the municipal parliament and the local tenants association (see following sections). Their decision based on the argument that "the introduction of a special land use plan would be equal to a restriction of private ownership. Such restriction of property rights, however, would be disproportionate and therefore illegal" (Zurich City Council, 2019b: 3). Finally, on March 10, 2020, the city council fully approved CS's building permit for Brunaupark (Huber, 2020).

Basel-City Government

In the year 2018, Basel-City has started its political debates on the revision of the Local Zoning Act.[12] The city council aimed to introduce "planning measures which lead to a density increase at inner-city locations to promote housing space for additional 5000 residents under the paradigm of green energy consumption" (Basel-City Council, 2018: 1). To meet this goal, the local government has introduced a progressive housing policy strategy. In practice, this means that the city council (executive) tries to purchase land for public housing and has expanded its

[12] Basel-City Local Zoning Act of 17 November 1999 (LZA 730.100).

collaboration with non-profit housing associations to increase the share of affordable housing (Basel-City Council, 2016: 38). Furthermore, the municipal government provides demand-side subsidies to low-income residents (Basel-City Council, 2016: 38).

> In addition to demand-side subsidies, the city of Basel has increased its supply-side subsidies to support the city's non-profit and affordable housing supply. For instance, the city provides long-term building leases on public land to non-profit cooperatives. (Interviewee 41, City of Basel, Head of Housing and Urban Development Department, August 20, 2019)

In Basel Schorenweg, however, the above-mentioned local planning and housing policy measures have not succeeded in preserving the social qualities of the area. Tenants are being dismissed even though the city council has tried to purchase the land in Basel Schorenweg for the provision of social housing units. In fact, they could not accomplish the purchase as CS's price request was too high for the city government (Interviewee 43, Credit Suisse portfolio manager Basel, September 12, 2019). In spring 2020, the city council granted the building permit and rejected the objections submitted by residents and the local tenants association. The decision was legitimized by the argument that Schorenweg is in private property and the densification measures announced by CS would take place within the regular Local Zoning Act.

Credit Suisse's Development Strategy

In Zurich Brunaupark and Basel Schorenweg, Credit Suisse has decided to densify the existing housing stock as both areas are centrally located. The possibility to rise density stipulated by the Local Zoning Acts has opened attractive investment conditions in both cities for them (Interviewees 39 and 43, CS portfolio managers in Zurich, July 8, 2019 and Basel, September 12, 2019).

> We decided to create more housing units at the same location. For us as pension fund this was all the more interesting. Because: where can one invest money nowadays anymore? We prefer to invest money at central locations which are well connected to transport nodes. (Interviewee 39, Credit Suisse portfolio manager Zurich, July 8, 2019)

In Basel Schorenweg, for instance, by 2040, it is estimated that the area will grow by +1000 new housing units. In 2009, the local planning authority has authorized the construction of two new housing skyscrapers in the area as well as a new school (Basel-City Council, 2009). By the end of 2018, these two high-rise buildings were finished (Oppliger, 2016). In March 2019, CS then decided to densify Schorenweg area as they aimed to benefit from these improved urban development and asset conditions (Interviewee 43, Credit Suisse portfolio manager Basel, September 12, 2019).

> Schoren area is an attractive city neighborhood with leisure and green facilities nearby. The area has been strongly developed in recent years. [...] However, even if the area would not have been developed, it would still be Basel center and attractive for investment. The area's upgrading clearly had a positive effect. (Interviewee 43, Credit Suisse portfolio manager Basel, September 12, 2019).

Even though CS in both settlements was aware that they renovated the buildings only a few years ago and a lot of criticism against the planning procedure was raised, they aimed to modernize the apartments to invest money at central location. The overall aim was to benefit from high return of investment through increasing rents at an attractive urban place.

> It is a fact that the rents will increase. [...] In the end, however, the buildings are newly renovated with less financial expenses for maintenance costs. For us, as investors with a long-term perspective, this is important. (Interviewee 43, Credit Suisse portfolio manager Basel, September 12, 2019)

Interestingly, in Zurich Brunaupark, CS states that they initially planned to densify via partial redevelopment rather than total reconstruction to protect social qualities (Interviewee 39, CS portfolio manager Zurich, July 8, 2019). This proposal, however, was disapproved by the Local Building Committee. As a consequence, CS decided to dismiss the residents and to demolish the existing housing stock to be able to build a totally new and architectonically homogenous settlement.

> The local building committee told us that Brunaupark settlement is too heterogenous. [...] Therefore, we changed our decision and aimed to perform a more homogenous and uniformed area redevelopment. (Interviewee 39, Credit Suisse portfolio manager Zurich, July 8, 2019)

Tenants' and Local Ngos Resistance Strategies

As a result of the socio-economic challenges tenants are confronted with due to densification and upgrading, in Zurich Brunaupark, on March 12, 2019, a group of tenants founded a local self-help initiative to make their protest visible. Today, "Brunaupark tenants association [BTA]" counts around 120 members and aims "to secure stable, affordable and socially-mixed housing for all income segments in the settlement" (BTA, 2019a). The association has organized street rallies and initiated a local petition which over 5700 citizens signed within one month (Interviewee 37, Head of Brunaupark Tenants Association, June 13, 2019). Specifically, the local petition called for a legal rejection of the contract terminations (BTA, 2019b). Simultaneously, left-wing parties of the municipal parliament (strongly supported by the local tenants association) initiated a referendum which aimed to introduce a "special land use zone"[13] for Brunaupark area. Legally, the introduction of such a zone would have put the municipal parliament in charge of approving development projects rather than the city council only (Zurich Municipal Parliament, 2019).

Similarly as in Zurich, in Basel Schorenweg, in March 2019, 96 households submitted a lawsuit against unfair contract dismissal to make their rights visible (Interviewee 38, head of Local Tenants Association, June 26, 2019). Moreover, in June 2019, Leilani Farha, UN housing expert, visited the settlements of Zurich Brunaupark and Basel Schorenweg as part of her Europe tour. To help the residents, she wrote an advisory letter to the Swiss Federal Office for Foreign Affairs on behalf of UN. Main topic of the letter was the unfair treatment of tenants, particularly, in regard to their social eviction at short notice and the precarious housing situation for old-aged and low-income households (Sturzenegger, 2020). As we have described above, however, none of these attempts were supported by the city authorities in Zurich or Basel.

### 5.3.5 *Explaining the Mechanisms at Play Leading to Residents' Social Exclusion in Zurich Brunaupark and Basel Schorenweg*

In this article, *first*, our goal was to explain how tenants are affected by densification from a social sustainability perspective and to understand how their positions are integrated into local decision-making. *Second*, we

[13] Special land use zones are designated to areas of increased public interest in which spatial development can take place outside the regular zonig plan.

analyzed the strategies developed by local authorities, property owners, and local NGOs to defend their interests in each institutional setting. This allows us to explain how and why trade-offs between environmental, economic, and social goals of densification take place. More specifically, we could highlight the institutional mechanisms at play leading to tenants' social exclusion in urban densification projects.

Our analysis reveals that in Zurich Brunaupark and Basel Schorenweg, municipal authorities approve CS' plans since the investor acts within the regular zoning plan. According to the protection guaranteed by the Swiss Constitution, no legitimate reason for private housing property restriction exists. In other words, in both cities, CS acts within the limits of the law—a law that is not targeting the protection of tenants' interests—and the authorities in charge of granting building permits do not try to interfere. CS hence seeks to benefit of the densification potentials guaranteed by the Local Zoning Act. The bank acknowledges densification as lucrative business as mortgage rates are low at the moment and the pressure to invest capital is increasing. Under the premise that the demand for housing will continue to be high, investment risks are minimal and urban densification is all the more profitable. Investment costs, in turn, can be amortized in very short-term (Aalbers, 2017; Brenner et al., 2012; Harvey, 2012; Marcuse, 2016). As a result, under the "flag" of densification as a public policy goal, CS realizes redevelopment projects with high return of investment. As long as they are not legally restricted to do so, neither in Zurich nor Basel, CS will feel compelled to support tenants' social inclusion.

To counteract private investors' development practices, municipal authorities have started to intervene more proactively into housing development in Zurich and Basel. For example, city authorities in both municipalities have introduced new housing policy measures such as increased subsidies for social housing associations. Moreover, local planners have also worked with property rights. In Basel Schorenweg, for example, the planning administration tried to purchase the private plot in Schorenweg but the price requested by CS was deemed too high. To legitimize such acquisition strategy, broad political support is needed which is not always given, even in big and wealthy cities such as Basel or Zurich. As a consequence, city authorities have neglected the social dimension of densification in Zurich Brunau and Basel Schoren.

To be more precise: in contrast to social objectives, which investors and municipal authorities perceive as a barrier to the economic development of cities, ecologic goals of densification bring investment opportunities, which explains why both parties are willing to agree with the new constraints at the expense of its social side. In fact, social inquiries make planning procedures more expensive for investors and public actors and potentially prevent the comprehensive urban renewal projects that planning administrations are supporting (see Brunaupark case). As a result, the Swiss legal context characterized by strongly protected property rights and weak tenancy law has led to a situation where the real-estate industry and municipal authorities work hand in hand to promote densification as "Eco-Business" at the expense of its social dimension.

As a consequence, our results show that long-term residents in Brunaupark and Schorenweg find themselves evicted as they are no longer able to afford the new rents of their modernized and densified dwellings. They are forced to leave urban centers because of the lack of affordable alternatives. This process of social exclusion contributes to the constant erosion of social relations and contacts to family members, neighbors, and friends. In the Brunau and Schoren areas, residents feel that their perspectives as tenants have not been respected by municipal authorities, particularly, since their legal objections (e.g. against unfair dismissal) have been rejected in both cities. Even though they tried to resist through street rallies, formal petitions, or the collaboration with the local tenants association, they did not succeed in defending their interests and faced discrimination because of their low-income status.

### 5.3.6 *The Eco-Business of Densification and the Neglection of Its Social Dimension*

Even though densification has become a core objective of urban policy agendas across the globe, critical analysis of its socio-political limitations, challenges, and contradictions, particularly in regard to its effects on tenants from a social sustainability perspective is largely missing (Ancell & Thompson-Fawcett, 2008; Burton, 2003; Pérez, 2020; Scally & Tighe, 2015). This article addresses this gap in the literature and focuses on the social implications of densification, therefore indirectly contributing to understand how the social qualities of a city can be sustained effectively.

Based on two Swiss examples, we show that the current way of implementing densification objectives can be far from socially sustainable. A powerful coalition between private landowners and municipal authorities promote densification as an "Eco-Business" by coupling urban competitiveness with ecologic viability goals, while neglecting social aspects. This coalition tends to jeopardize the very social qualities which are a city's basis of community-based initiatives and solidarity-creating capacities. Low-income groups, including old-aged, young families, or student households, find themselves in a vicious circle in which they do not have other alternatives than to move to cheaper suburban areas. Hence, those being displaced have become the victims of powerful forces of capitalist urbanization and differential spending power.

Even though our results are limited to two Swiss cities, potential for generalization to other urban contexts results from the identified causal mechanisms which explain why the social dimension of sustainability tends to be bypassed in densification processes. On this basis, we can formulate different strategies to make densification more social (e.g. quota for affordable housing, public subsidies for non-profit housing cooperatives, eviction controls, etc.). An effort can also be made to properly activate existing instruments: making densification more social is not only a matter of political will, but also of the ability of public administration to familiarize with all the range of existing intervention possibilities. Sometimes the implementation of more social measures might have to be done at the expense of architectural quality and homogeneity, but there might be good reasons to do so. Residents could also be involved more actively into decision-making and stricter control mechanisms in relation to occupancy rate rules or income levels in social housing units could be activated. Greater awareness of the detrimental social implications of densification and implementation of proactive measures to counteract them could also improve the acceptance of densification and prevent NIMBY-responses (Scally & Tighe, 2015).

The findings of this article call also for a greater involvement of city governments to counteract trends of increasing commodification of urban housing stocks (Aalbers, 2017; Marcuse, 2016; Rolnik, 2013) by systematically promoting new and/or available policy measures which incorporate how residents interpret the places in which they live and embedded in order to include their perspectives into local decision-making procedures and regulations. Only in doing so, cities will succeed in promoting densification projects that are designed to the needs and capabilities of people who are actually supposed to benefit from them—the residents.

Presumably, sustainable urban development calls for economic stability, environmental protection, and social sustainability not just the one or the other.

## 5.4 Article 3: Strategic Use of Land Policy Instruments for Affordable Housing—Coping with Social Challenges Under Scarce Land Conditions in Swiss Cities

Gabriela Debrunner & Thomas Hartmann
Journal: Land Use Policy, 2020, Vol. 99, https://doi.org/10.1016/j.landusepol.2020.104993
Impact Factor (2019): 3.682

Status: Published

**Abstract:** For decades, many cities have introduced densification policy objectives to stop urban sprawl or to promote efficient use of natural resources. In the urban housing sector, however, densification projects often intensify social challenges. Due to rising rents after modernization of existing housing stocks as a consequence of densification, low-income tenants are forced to leave their apartments. Risks of social exclusion and segregation increase simultaneously.
In this paper, we analyze how local planning authorities cope with affordable housing shortages in a context of urban densification. Specifically, we ask: *How do municipal planning authorities promote affordable housing in densifying cities?* To answer this research question, we apply a neoinstitutional analysis approach to better understand (1) the basic mechanisms of how land policy instruments impact affordability, and (2) why specific instruments are activated to defend affordable housing objectives. Through qualitative case study analysis of four Swiss urban municipalities, our results show that the mere availability of land policy instruments is not sufficient but that the strategic activation of specific instruments matters.

*Keywords: land policy; affordable housing; densification; gentrification*

**Research Highlights:**
*The paper...*

- *... applies a comparative case study approach to analyze the municipal land policy strategies for affordable housing in four Swiss cities.*
- *... analyzes how Swiss municipal authorities in charge of land-use planning promote affordable housing in a context of densification.*
- *... explains what kind of policy instruments they can use to promote affordable housing effectively.*
- *... argues that effective municipal land policy in favor of housing affordability (as key element of urban social sustainability) does not only involve the mere introduction of new instrument.*
- *...but the strategic activation of available instruments matters.*

### *5.4.1 The Role of Municipal Planning Authorities in Affordable Housing Provision*

For decades, many cities have introduced densification policy objectives to stop urban sprawl or to promote efficient use of natural resources (Daneshpour & Shakibamanesh, 2011; Touati-Morel, 2015). Densification is defined as a process leading to higher exploitation (number of households) within existing city boundaries (Boyko & Cooper, 2011: 47). The implementation of this process, however, is a contested procedure as it requires to deal with the already built environment (Gerber et al., 2018). In the urban housing sector, for example, the compact city may improve public transport use, while at the same time it is likely to mean less domestic living space and a lack of affordable housing. A rising number of tenants suffer from social eviction, contract termination and dismissal due rent increase after modernization as a direct consequence of urban densification (Bramley et al., 2009; Burton, 2003; Chiu, 2003).

In Switzerland, for instance, a legally binding densification policy objective has been introduced in the revised Federal Spatial Planning Act (SPA) in 2013. Swiss municipalities are obliged to densify through inward settlement development to protect agricultural land and to preserve natural habitats (Art. 1 SPA). However, since the majority of Swiss cities lacks of free inner-city plots, the process is primarily being implemented through redevelopment of existing housing stocks e.g. in the form of total replacement construction, renovation, or compaction of existing buildings (Nebel et al., 2017). Simultaneously, affordable housing shortages for low- and middle-income households have intensified in almost every Swiss city in recent years (FOH, 2016a). The current housing situation in urban areas is charaterized by an overheated housing market with vacancy rates below 1% and rising rents (Balmer & Gerber, 2017). Consequently, municipal authorities are increasingly confronted to deal with tenants suffering from social exclusion due to rising rents after modernization and densification. Particularly, low-income old-aged, migrant and family households are affected from social displacement as newly renovated buildings are only affordable for middle and high-income groups and non-profit housing suppliers have long waiting lists (FOH, 2016a). The current situation is even more worrying considering that Switzerland is regarded as a nation of tenants with having the lowest homeownership rate in Europe (Lawson, 2009). Housing provision traditionally lies in the responsibility of the profit-oriented private rental sector (Bourassa et al., 2010). In Swiss cities (where over 70% of the population lives), 63% of households live in apartments of private investors and are strongly dependent on the housing stock owners' decisions (FOH, 2017).

In this article, we analyze how municipal planning administrations cope with affordable housing shortages in a context of urban densification. Specifically, we explain the strategies implemented by municipal authorities in order to deal with affordable housing provision when confronted with scarcity of land. We ask: *How do municipal planning authorities promote affordable housing in densifying cities?* To answer this research question, we apply a neoinstitutionalist analysis approach and proceed in two steps to better understand (1) the basic mechanisms of how policy instruments impact affordability, and (2) why local planners activate specific instruments to defend housing affordability objectives. These questions require the use of qualitative case study methodology (Yin, 2018) employed in four Swiss municipalities which are all confronted with rising affordable housing scarcity and densification pressure. This article brings together examples of local practices to raise awareness how planners can strategically activate different policy instruments to promote affordable housing in urban areas.

### 5.4.2 *Affordable Housing Provision at the Interface Between Public Policy Intervention and Property Rights*

Housing affordability refers to a situation in which households are able to pay a certain percentage of their income for housing costs so that they will have enough left for other necessities of life (Stone, 2006). In Switzerland, for instance, it is generally assumed that a quarter (25%) of the monthly net household income can be spent on the monthly gross rent without negatively impacting other dimensions of life (FOH, 2014c).[14] To understand the diverse mechanisms that influence the status of housing affordability, we apply a neoinstitutionalist analysis approach (Gerber et al., 2009; Mandelbaum, 1985; Ostrom, 2007; Williamson, 2000). In general, this approach focuses on the relationship between institutions and actors' decisions observing that human actions take place within a tight web of formal rules which structure individuals' expectations about what others will do (Hall & Taylor, 1996: 956). Institutions are thereby understood as shared social values stipulated in formal laws and ordinances which guide social interaction and practices (Dembski & Salet, 2010: 612). Following the neoinstitutional perspective, housing affordability is hence regarded as the result of human actions. The key attributes of individuals' behavior—both the institutional setting and the actors' self-interestedness and strategies of action—need to be addressed for understanding the (un)sustainable use of urban housing stocks (Williamson, 2000: 600).

*Strategic Activation of Land Policy Instruments for Affordable Housing Provision*

Housing is a resource that is economically significant. Traded on the free market, housing is a commodity with enormous economic potential which is why it is often treated as a highly valued collateral. Especially in cities, where demand for housing is high and the potential for capital accumulation is lucrative, the competition between actors interested in using urban land for housing is rising and rents constantly increase (Aalbers, 2017: 543–544). In addition, in many cities, the prohibition to build outside municipal boundaries and the obligation to densify within the built environment has even reinforced this competition. Due to scarce land resources, the prices for land and the housing stocks which are built on intensively increase (Burton, 2000: 1976).

Consequently, the provision of affordable housing in dense city areas is to be regarded as a land policy issue since owners can ask for higher land

[14] In comparison to other European countries (e.g. Germany) where normally a 30% share is regarded reasonable, in Switzerland, monthly fix costs such as taxes (approx. 10% of the monthly loan) and health insurances (approx. 250 CHF per month) are not already deducted from the monthly income. Therefore, a quarter is considered to be appropriate.

prices due to increased proximity within the financial center, centrality to transport nodes, or accessibility to services which directly influences the rental cost level (Theurillat et al., 2014: 1426). In this context, "land policy" encompasses all the political-legal measures implemented by the municipality to deal with the issue of land use regulation (Hartmann & Spit, 2015). More precisely, land policy "is defined as all those state decisions and measures that have an influence on the way land is used, distributed and valued to implement the politically defined spatial development goal" (Hengstermann & Gerber, 2015: 246). This active definition of land policy (see for discussion, Hartmann & Spit, 2015; Healey & Barrett, 1985; Knoepfel et al., 2012; Needham & Verhage, 1998; van der Krabben & Jacobs, 2013) differs from a passive land policy understanding (e.g. Davy, 2005: 117) that focuses not only whether land is changed (passive) but also in regard to achieve a specific spatial development goal (e.g. affordable housing provision) (active) (Hengstermann, 2019).

In practice, such state interventions appear in the form of public and private law instruments that operate according to a different logic and rely on different forms of legitimacy (Hood & Margetts, 2007; Needham et al., 2018). In general, policy instruments are defined as intervention ways or measures that are needed to achieve a certain public policy goal (Knoepfel et al., 2007: 156–157). To provide affordable housing, for instance, different intervention ways for municipalities exist including the increase of social housing (public ownership) or a shift towards project-based subsidies that may stimulate the affordability of rents (Kadi & Ronald, 2014: 271).

- *Public law instruments* derive from public policy including regulatory statues, penal laws, and other laws of public order. They aim to solve a political problem that was defined as such by the voting majority e.g. urban sprawl. Public law instruments such as subsidies or zoning measures are regularly revised, not only because the problem they are targeting constantly evolves, but also because changing political majorities propose alternative solutions to the problem (Knoepfel et al., 2012).
- *Private law instruments* derive from private law including property law, the law of contracts, torts and obligations. Their aim is to defend private interests against the (potentially absolutist) power of the state (Locke, 1689). Property rights are grounded in the Civil

> Code (or similar in common law contexts) and are extremely stable over time because their definition hardly changes (Bromley, 1992; Savini et al., 2015). Without heavy state intervention such as expropriation, therefore, new planning regulations only get implemented when titleholders agree to undertake new development, sell their land, or transfer their development rights (Gerber et al., 2017).

Under scarce land conditions, city authorities frequently fail to cope with complex property-right arrangements as most instruments were crafted to deal with use situations on unbuilt greenfield. Densification, in contrast, implies to deal with the already built environment and with complex property situations (e.g. small-scale ownership, veto rights controlled by power actors). Therefore, in a densifying city, planning for affordable housing requires a keen understanding of the instruments available to govern the close interactions between land use planning (public policy) and property rights (Blomquist, 2012; Dawkins & Nelson, 2002). Planners have to understand that the selection and combination of policy instruments is never neutral. Rather the choice corresponds to a specific interpretation of the role played by the state and/or its private partners (Salamon, 2000). An active land policy strategy, in other words, requires planning administrations, which are capable to develop intervention ways to reinforce their position in front of powerful landowners or economic interests to address complex and rivalrous land use situations (Alterman, 1990). These strategies are deliberate and goal-oriented applications of certain instruments or the combination thereof (Gerber et al., 2018). For example, as densification implies a form of planning that goes beyond zoning in order to deal with complex property rights situations, strategic competencies of municipal authorities include e.g. amicable purchase, building leases, public–private-contracts, and so on. Moreover, because property titles give additional power to their holders to shape housing development, public actors too can use them to reinforce their position (Gerber et al., 2017: 1687).

### *5.4.3 Comparative Case Study Analysis of Four Swiss Cities*

To investigate a contemporary phenomenon in-depth—the provision of affordable housing in densifying urban areas—and as a results of human action, the empirical material of this study is conducted through qualitative case study methodology (Yin, 2018).

*Case Selection and Methods*

The study is conducted in two steps: in a *first step*, we investigate the wide range of available policy instruments to promote affordable housing in cities under densification pressure (sub-question 1). Following this objective, Switzerland makes an interesting case study to analyze the relationship between land use regulation, densification, and housing affordability as the challenge of coordinating the three has become predominant in the country in recent years (Gennaio et al., 2009; Rérat, 2012; FOH, 2016, 2016a). Especially since the approved revision of the Federal Planning Act in 2013 which enforces the 26 cantons and over 2000 municipalities to promote "inward settlement development, while ensuring an appropriate quality of housing" (Art. 1, para. 2, lit. abis SPA). Simultaneously, population growth coupled with yield-oriented investments attracted by the state's economic stability and wealth reinforced the attractiveness of Swiss real-estate markets. Triggered by low-interest rates, urban housing has become the main target of capital investment, especially for pension funds. As a result of increased commodification, modernization, and densification of housing stocks, the provision of affordable housing has increasingly become difficult for Swiss planning authorities in recent years (Balmer & Gerber, 2017).

The country is organized on three executive levels—municipalities, cantons, and the confederation—and is characterized by a form of "cooperative federalism". Local zoning plans are prepared by municipal authorities but must comply with cantonal and federal plans (Linder, 1994). In any case, local zoning regulation is binding to private property owners. Investors and developers can be asked to prepare and to fund specific plans, however, the final validation of the local plan and the provision of building permits always relies in the responsibility of the municipal planning authority (Gerber et al., 2017: 1690). Methodologically, in this step, we performed a broad screening and analysis of policy documents at the national and local level. We analyzed articles published within the last decade, including government reports, vote results, legal documents, parliamentary debates, newspaper articles, and "grey" literature which documented the use and range of available policy instruments for affordable housing provision. Our goal was to detect the wide range of policy instruments available for Swiss municipalities.

In a *second step*, we analyzed the municipal authorities' strategies when activating specific policy instruments for affordable housing provision (sub-question 2). Therefore, we selected two cases in urban core areas—the cities of Zurich and Basel—and two cases in suburban areas—the cities of Köniz and Kloten—to gain a broad understanding of the strategies applied in different urban contexts. Whereas the cities of Zurich and Basel face a period of severe urban housing shortage, in the cities of Köniz and Kloten the pressure on affordable housing provision is not as profound but is constantly rising. In Zurich, for instance, rental costs on the private rental market have increased by +75% since 2000 (Balmer & Gerber, 2017: 8). On average, a $90m^2$ apartment offered on the rental market costs more than 2000 Swiss francs net-rent per month which is only affordable for middle and high-income households (Schmid, 2020). Hence, moving to cheaper suburban areas has remained the only option for many vulnerable and lower-income groups in Zurich.

The four cases were selected as they all show similar socio-economic characteristics (population growth, vacancy rate, land scarcity, densification pressure, affordable housing shortage) and, at the time of investigation, had to deal with ongoing densification projects in the urban housing sector (Table 4.1). Thereby, we were able to directly confront the actors involved with the decisions taken in relation to affordable housing provision and densification. To achieve this goal, we conducted ten semi-structured interviews with eight experts from the public sector (local planning, housing, and social welfare departments) and two local tenants associations. In addition, we analyzed each case 20–30 policy and project documents to gain a detailed understanding of the applied municipal authorities' interests and strategies.

*(see Table 4.1 of this book)*

### 5.4.4 *Strategic Use of Land Policy Instruments for Affordable Housing*

As mentioned before, municipal authorities may activate different policy instruments to alter land parcels in size and shape in order to promote affordable housing. In doing so, they aim to change the use conditions for specific groups. In the following section, we present four strategies which Swiss municipal authorities follow to promote affordable housing (Fig. 5.3). These intervention ways derive from the neoinstitutional analysis approach (Gerber et al., 2018) and structure both the empirical analysis and the discussion section of this paper. The four strategies were selected along their characteristics (either deriving from public or private law), and their potential to defend municipal interests in front of powerful landowners. In particular, we distinguish between, first, instruments that regulate land uses by using public policy with no direct impact on the use rights of land such as economic incentives for landowners (see [1] *supply-subsidies*). Second, instruments using public policy leading to a regulation of use rights on formal ownership ([2] *zoning*). Third, instruments leading to a legal redefinition of property rights in the Civil Code ([3] *contracts*). And forth, instruments that redistribute property rights such as expropriation or targeted purchase of land ([4] *property rights*).

For each policy instrument, we briefly explain how they work in the Swiss context. Then, based on our qualitative case study analysis, we investigate how the policy instruments are used for affordable housing provision in the four cities. Third, we explain why local planners activate specific policy instruments to promote affordable housing in order to understand the strategy behind the policy measures applied. As the process is ongoing, the list is neither to be evaluated as complete nor exhaustive.

*The Basic Mechanisms How Policy Instruments Work in the Swiss Context*

(1) Supply-Side Subsidies

In general, the granting of supply-side subsidies does not target the individual tenants, rather individual buildings. In the housing sector, Swiss municipalities provide supply-side subsidies in the format of economic incentives such as direct loans, tax relief, bank guarantees, or advantageous mortgages to private third parties e.g. to non-profit associations or private individuals.

*For non-profit housing associations*: Private homeowners have to belong to a federal umbrella organization to benefit from public supply-side subsidies for housing purposes. The organization controls that its members (mainly non-profit cooperatives and foundations) provide affordable housing according to the cost-rent principle and based on non-profit objectives only (Balmer & Gerber, 2017).

*For private individuals*: Moreover, each Swiss city (based on the Federal Energy Act) does not only provide supply-side subsidies to the non-profit housing sector. They are also obliged to grant subsidies in the form of direct grants, financial incentives, tax relief, and free consulting to private individuals (including institutional investors) to improve energy

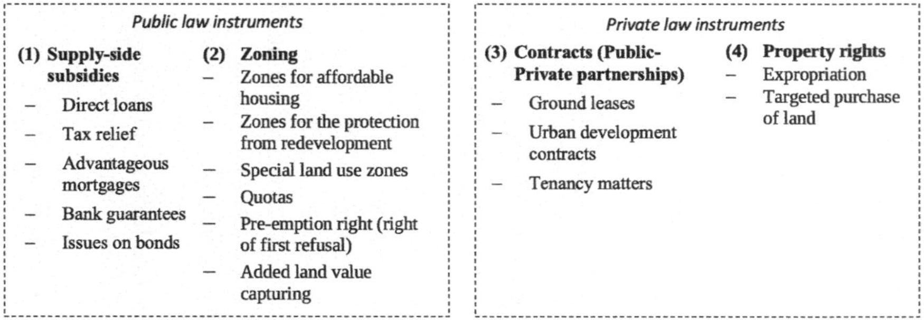


**Fig. 5.3** Municipal land policy instruments for affordable housing in Swiss municipalities (analysis based on e.g. FOE & FOH, 2015; FOE, 2015, 2018; FOF et al., 2009; FOH 2012, 2014, 2014a, 2014b, 2016, 2016a, 2017, 2018b, 2019; FOSD, 2016; FOSI & FOH, 2015; Swiss Federal Council, 2014, 2016a, 2017)

efficiency standards e.g. for insulation, windows, or heating. So far, the granting of energy subsidies for private individuals has only been determined by technical criteria and has not been attached to the fulfilling of social objectives. However, in recent years, the federal government has started to investigate whether such granting of subsidies could be coupled to social tasks too e.g. to the requirement to provide affordable housing if one aims to benefit of subsidies (FOH, 2016).

(2) Zoning

In Swiss municipalities, zoning is a relatively new land policy instrument for the provision of affordable housing. However, zoning measures have gained in strategic relevance to steer affordable housing development in recent years since urban land has become scarce and increased flexibility is needed (FOH, 2012). The policy instrument allows municipal authorities to directly intervene into private development plans because zoning regulations are binding to landowners.

*Zones for affordable housing*: In these zones, the municipality can oblige private landowners to provide a minimum share of affordable housing (e.g. 50%). Thereby, the amount of affordable housing can be raised effectively since all private landowners are legally obliged to follow this objective in these zones.

*Zones for the protection from redevelopment*: Landowners only receive a building permit to (re-) develop, renovate, modernize, or replace existing housing stocks in these zones if tenants will have the opportunity to stay in their apartments afterwards. For example, property owners must approve that the rents will not exceed a certain level up to three years after the renovation task is finished. The primary political objective behind this measure is to keep the rents low when demand is high and to prevent luxury renovations.

*Special land use zones*: The introduction of special land use zones is designated to areas of increased public interest in which municipal planning authorities can encourage spatial development outside the regular zoning plan. These zones are legally binding for public authorities and landowners but the private parcel's development terms and conditions are still negotiable for both sides e.g. regarding use density requirements, energy efficiency standards, urban design or housing affordability objectives. So far, in many Swiss cities the instrument has primarily been used for unbuilt industrial zones (e.g. for the transformation of former train station areas) as well as for greenery and infrastructure projects (e.g.

lake shores, hospital areas, education facilities). In the housing sector, however, the instrument has gained new strategic relevance too since local planners are obliged to increase density within municipal boundaries and increased flexibility is needed (Knoepfel et al., 2012: 423). Particularly, through the use of this instrument, city governments try to couple density goals with housing affordability objectives. For example, in these zones, the municipality can oblige the private investor to provide at least 40% affordable apartments, and in turn, landowners can benefit of a density increase of +10% outside the regular zoning plan.

*Quotas*: The introduction of quotas for affordable housing is a quantitative zoning mechanism which assures municipalities that the ratio between low-cost housing supply and demand does not exceed a certain level. Through the use of quotas, planning authorities can steer affordable housing provision effectively as the output is regularly controlled and monitored by quantitative guidelines. In particular, quotas help planners to have a clear vision and goal what type and size of housing units need to be built within a specific timeframe e.g. by 2050. The instrument also helps city authorities to legitimize the use and introduction of additional policy instruments (e.g. the purchase of private land) which also support the increase of affordable housing in the long term.

*Added land value capturing*: The policy instrument of added land value capturing is a zoning mechanism with which municipal governments reap some of the increment in land value attributable to planning decisions (Alterman, 2012). According to the revised Federal SPA, Swiss municipalities are obliged to capture a minimal taxation rate of 20% of the added land value for new-built housing on unbuilt land. The tax is due when the land is developed or sold (Art. 5, par. 1 SPA). Optionally, municipalities can also capture added land values that occur through densification measures on already built land (up-zoning) (Viallon, 2018). The funds collected by the instrument grant municipal authorities the possibility to distribute, remove, and relocate private development rights according to societal needs. For example, municipalities can capture a minimum share (e.g. 40%) of added land values that evoke through planning measures for affordable housing provision.

*Pre-emption rights*: A public pre-emption right (stipulated in the Local Zoning Act) assures the municipality the right of first refusal when private property is sold. Planners apply the instrument for the construction of, among other purposes, social housing units. In planning practice, the municipality makes use of a pre-emption-right if a private parcel has strategic relevance for the city's urban development as a whole e.g. for the construction of schools or to intervene into socially segregated areas. It also provides municipalities the capacity to prevent land speculation which could hamper affordable housing provision in general (Nahrath, 2018).

(3) Contracts

According to Swiss private law (see Swiss Civil Code and Federal Obligations Code), a contract is defined as a legal agreement between two or more parties, enforceable by law, to perform a specified act. In case one of the two is a public actor, the contract is to be considered as "public–private-partnership" (Nicol & Knoepfel, 2008: 172).

*Long-term ground leases*: Ground leases grant the landowner the right to retain legal ownership while transferring the right to use his/her land to a private third party (Gerber, 2018). In Switzerland, ground leases are granted for up to 100 years in exchange for annual rent payment. At the end of the lease period, all improvements made to the land by the owner of the building revert back to the landowner, according to the terms of the initial contract (Gerber et al., 2017: 1690). With regard to affordable housing provision, Swiss cities often use the instrument for collaboration with non-profit housing associations (Balmer & Gerber, 2017). In practice, the municipality remains the landowner while the ownership of the building is transferred to a private third party such as a non-profit housing cooperative. The municipality as landowner benefits from a stable source of income over time through lease revenue but does not bear the financial risks to manage the use of the building. The municipality remains in charge to determine special use requirements on their plots e.g. related to social mix rules, income levels, or housing affordability objectives.

*Urban development contracts*: In urban development contracts, another form of "planning by contract", a private landowner aims to improve the use requirements that are set on his/her private parcel in the regular zoning plan through a renegotiation of the terms and conditions with the municipal planning authority. Planners may agree to such (re)negotiations with the private investor, for instance, to promote affordable housing. Under scarce land conditions, in particular, the municipality prefers to influence the development terms on private land on soft scale rather than not having an impact on private land at all. Through development contracts, planners can change the "rule of the game" quickly, flexibly, and for specific private locations only (Feldges, 2019).

*Tenancy matters*: Ultimately, the support of affordable housing can also be promoted via tenancy law. Swiss tenants' rights are protected by the articles for tenancy matters in the Federal Obligations Code (Art. 253–274 OC) as fifth part of the Swiss Civil Code. In international comparison, tenants in Switzerland are considered to be weakly protected by law in relation to neighboring states (e.g. Austria, Germany) (GFOBRP, 2016). For instance, landowners are allowed to terminate an open-ended rent contract within three months without any specific reason. So regardless of the tenants' strength of social integration, age, or years of residency in the neighborhood. Residents do also not need to be informed about upcoming densification tasks before receiving the contract termination which leads to social eviction at short notice. Tenants may counteract in court, however, in most cases they do not use this option as they neither have the financial means nor the expert knowledge to do so. Therefore, some Swiss cantons (e.g. Zurich, Fribourg, Basel-City, Geneva) have revised its Cantonal Tenancy Act to better protect tenants from rent increase and dismissal (FOH, 2018a).

(4) Property Rights

*Expropriation*: In Switzerland, the "right to own" property is protected as a fundamental right by the Swiss Constitution (Art. 22ter CSC). As such it can only be restricted if (1) a legal basis and an overweighting public interest exist; (2) the measure is proportional; and (3) a full compensation is paid (Art. 5, Art. 36 para. 1–3 & Art. 26 para. CSC). Especially in the housing sector, Swiss courts interpret the weight of public interest narrowly so that property restrictions are limited and expropriations are rare in international comparison (Alterman, 2010). As a consequence of this legal situation, the political legitimacy to make use of expropriation for affordable housing provision is low as expropriation is politically contested, long, and expensive.

*Targeted purchase of land*: The targeted purchase of land guarantees the municipality the full right of disposal and the power to grant the land use rights on their plots. Through public ownership, the city as landowner is in charge to develop housing according to public preferences e.g. to fulfill social objectives such as the provision of affordable housing, the protection of tenants' rights or to ensure residential stability of old-aged. However, in daily planning practice, the purchase of urban land is expensive and political majorities for the support of such acquisition strategy is not always given. Therefore, many Swiss municipalities have started to intervene into private developers rights in other ways, especially through zoning measures.

*The Use of the Policy Instruments for Affordable Housing Provision in Each City*

In the following section, we explain *how* municipal planning authorities in the four cases investigated used which land policy instruments for affordable housing provision. Thereby, we identify differences but also similarities in Swiss local affordable housing practice. The data of the table (Table 5.5) derive from intensive literature and policy documents review and was supplemented by qualitative interview data.

*Strategic Activation of Specific Policy Instruments to Defend Housing Affordability Objectives in Each City*

In the following section, we explain why municipal planning authorities in the four cities activate specific policy instruments to defend affordable housing objectives.

The Case of Zurich

To reach the Constitutional mandate of 33.3% social housing property by 2050, Zurich's municipal planning authority makes not only use of public law instruments (e.g. supply-side subsidies, zoning). But the city council also commits to find other ways to increase the share of affordable housing effectively, particularly, by activating private law instruments too (e.g. land acquisition, long-term ground leases, changes in tenancy matters). Overall, the quota introduced in the Local Constitution helps the city government to legitimize the activation of additional policy measures such as the purchase of private land even though such acquisition strategy is expensive and politically contested in the local legislative parliament.

> The city of Zurich is committed to promote affordable housing in all its neighborhoods through zoning measures, supply-side subsidies for non-profit cooperatives, and the purchase of land for public housing to reach the constitutional mandate of 33.3% non-profit housing property by 2050. (Zurich City Council in Regional Zoning Act[15] 2019: 109)

Moreover, to effectively control the quantitative output and the affordability performance of each publicly subsidized housing association,

[15] Regional Zoning Act (2019): Planning of Municipal Settlements, Landscape, Public Facilities and Spaces. Zurich. 1–166.

**Table 5.5** Use of the land policy instruments for affordable housing provision in each city

| | *Case of Zurich* |
|---|---|
| **(1) Supply-side subsidies**<br>For non-profit housing associations | Since 1907, the city of Zurich provides subsidies to non-profit coops, housing associations, and foundations directly or via municipal foundations (e.g. «Stiftung PWG», «Stiftung Einfach Wohnen») in the form of advantageous mortgages, direct financial grants, loans, and issues on bonds. The provision of subsidies is connected to strict require-ments related to the fulfilling of social and ecologic tasks e.g. income guide-lines, social-mix, and occupancy rate rules, early communication with tenants as well as obligations related to urban design, green, cultural, and open spaces, or efficient use of energy |
| For private individuals | The Canton of Zurich (to which the cities of Zurich and Kloten belong to), initiated funding programs for private individuals such as "starte" as well as advantageous mortgages for private renovations, and targeted tax relief for private individuals to promote modernization of private housing stocks but without obliging homeowners to pro-vide affordable housing |
| **(2) Zoning**<br>Special land use zones | So far, Zurich mainly has used the instrument for unbuilt industrial zones e.g. Neugasse, Zürich-West, Manegg In Neugasse, for example, the city has obliged the investor to provide min. 30% of the newly built apartments created through rezoning measures for affordable housing provision. The dwellings must align with the cost-rent principle and are reserved for social housing associations |
| Quotas | In 2011, the city of Zurich introduced a fix min. quota of 33.3% of non-profit housing property in the Local Constitution. Following the revised legislation, the city must approve that by 2050 a fix minimum share of 33.3% of the total housing stock will be in social housing property (Art. 2, par. 5.4) |

(continued)

**Table 5.5** (continued)

| | *Case of Zurich* |
|---|---|
| Added land value capturing | The Canton of Zurich (to which the cities of Zurich and Kloten belong to) has agreed to oblige municipalities to capture a minimum share of 40% of added land values that evoke through densification/planning measures on already built land for, among other purposes, affordable housing provision in May 2019 (see Art. 49b Cantonal Building and Planning Act). The revised legislation will come into force in January 2021. However, it is not yet clear how exactly each municipality will apply the instrument at the local level |
| Pre-emption rights | In the Canton of Zurich—based on the Cantonal Building and Planning Act (§118)—municipalities are allowed to make use of public pre-emption rights via local zoning. However, so far, municipalities such as Zurich or Kloten have not used this instrument for affordable housing provision as its implementation is politically contested |
| **(3) Contracts**<br>Long-term ground leases | The city of Zurich provides long-term ground leases on public land to non-profit housing associations and municipal foundations. At the end of 2018, the municipality has provided 220 lease contracts to non-profit associations in total. The provision of ground leases is connected to the fulfilling of social tasks e.g. housing affordability, social mixing, and occupancy rate rules |
| Urban development contract | So far, the city of Zurich has not made use of urban development contracts for affordable housing provision. In future terms, however, it is expected that the instrument will gain in relevance as planners will need to negotiate the terms and conditions of private development within municipal boundaries more intensively |

(continued)

**Table 5.5** (continued)

| | *Case of Zurich* |
|---|---|
| Tenancy matters | Since November 1st 2013, based on a revision of the Cantonal Tenancy Act (§229b), tenants living in municipalities of the Canton of Zurich (e.g. Zurich city, Kloten) can force the property owner to disclose the former rent if they enter a new rent contract. In case the new rent does not align with the current interest rate, tenants are allowed to claim the rent increase in the cantonal tenancy court. In practise, however, tenants do often not use this option as they do not have the knowledge or financial means to do so |
| **(4) Property rights**<br>Purchase of land | The city of Zurich actively purchases private land for public purposes (e.g. for the construction of low-cost housing, schools, health facilities). A current example is the purchase of the 'Hornbach' settlement next to the lake of Zurich where 125 new social housing units have been built on former private land |
| | *Case of Basel* |
| **(1) Supply-side subsidies**<br>For non-profit housing associations | Since 1900, Basel-city provides subsidies to non-profit housing associations and cooperatives in the form of state guarantees, issues on bonds, advantageous mortgages, direct loans, grants, and tax relief. Same as Zurich, Basel provides subsidies to non-profit housing cooperatives connected to social objectives. For example, housing cooperatives must provide a certain share of the housing stock to the lowest income segments or for social welfare recipients only |
| For private individuals | The City of Basel provides energy subsidies to private homeowners in the form of direct grants, tax relief and free consulting but without obliging private homeowners to fulfil social tasks (e.g. affordable housing) |

(continued)

**Table 5.5** (continued)

| | *Case of Basel* |
|---|---|
| **(2) Zoning**<br>Special land use zones | Basel-city primarily used this instrument for unbuilt industrial zones e.g. Dreispitz, Klybeck, Volta, Erlenmatt. In these zones, the city obliges the investor to provide e.g. up to 30% of the total housing stock created through rezoning measures for affordable housing |
| Quotas | Basel-city has not yet introduced a fix min. quota of affordable housing property. However, based on a revision of the Local Constitution on June 10th 2018, the city government aims to introduce such a quota of 25% non-profit housing property by 2050 in the Local Housing Act |
| Added land value capturing | The city of Basel has introduced a tax on added land value created through zoning measures in 1977 already. It applies a uniform of 50% tax rate on the difference between old and new land market values, both to new building zones and to up-zoning changes. So far, however, the funds collected are primarily used for greenery not for affordable housing purposes |
| Pre-emption rights | Basel-city does not make use of public pre-emption rights, neither on the cantonal nor municipal level. However, based on the approved local referendum on June 10th 2018, the city government now plans to initiate the introduction of such a right to increase the share of affordable housing in the long run |
| **(3) Contracts**<br>Long-term ground leases | The city of Basel provides long-term ground leases on public land to non-profit associations for residential, commercial, and creative use. With 680 lease contracts in 2019 in total, the city of Basel is one of the strongest ground lease providers in Switzerland. In the housing sector, the provision of ground leases is connected to the fulfilling of social tasks e.g. affor- dability- and social mix rules |

(continued)

**Table 5.5** (continued)

| | |
|---|---|
| | *Case of Basel* |
| Urban development contract | In Basel-city, the local planning administration uses the instrument for specific areas e.g. for the development of the Novartis campus. So far, however, it has not been used for affordable housing provision only, but mostly, to improve green and open spaces |
| Tenancy matters | In addition, in June 2018, Basel-City has revised its Local Constitution to better protect tenants from redevelopment and dismissal. Following the revised legislation, the municipality is obliged to ensure that people who live and are registered in Basel city can rent an apartment that suits their income adequately. The rental costs are not allowed to exceed the respective household income or financial capacity. Moreover, private homeowners only receive a building permit for renovations, replacements, and demolitions if rents after modernization do not exceed a certain level. The revised tenancy legislation, however, is not yet in force |
| **(4) Property rights** Purchase of land | The city of Basel purchases private land for public purposes e.g. for housing, schools, or health care. Since 2007, the net share of public property has constantly grown although not with an explicit focus on affordable housing provision |
| | *Case of Köniz* |
| **(1) Supply-side subsidies** For non-profit housing associations | In March 2017, the voting majority in Köniz agreed to expand supply-side subsidies for non-profit cooperatives in the form of advantageous mortgages, direct grants and loans so that non-profit coops can afford e.g. ground leases on public land or the purchase of private land. The provision of subsidies is also connected to social requirements such as income guidelines or social mix rules |

(continued)

**Table 5.5** (continued)

| | *Case of Köniz* |
|---|---|
| For private individuals | The City of Köniz provides energy subsidies to private individuals in the form of direct grants, tax relief and free consulting but without obliging private homeowners to promote affordable housing |
| **(2) Zoning**<br>Special land use zones | According to the revised Local Zoning Act, the city uses the instrument for areas with more than 4000m$^2$ floor area only. Here, the municipality obliges the investor to provide min. 20–40% of the newly built apartments created through the approved density increase for affordable housing provision. In practice, however, Köniz only has little areas that are big enough to suit this condition |
| Quotas | In the Canton of Bern, municipalities are obliged to fulfil densification objectives according to quantitative guidelines (quotas). For example, Bernese municipalities such as Köniz can introduce a fix minimum share of housing space consumption per person for specific locations e.g. for selected densification areas. Optionally, they are also allowed to couple density requirements with housing affordability objectives. However, in the city of Köniz none of these quantitative options are used for affordable housing provision |
| Added land value capturing | Since March 2020, Bernese municipalities are obliged to capture min. 20% to max. 50% of added land values created through densification and zoning measures on unbuilt land for public purposes of different kind (e.g. for affordable housing). Optionally, municipalities can capture added land values that occur through densification measures on already built land (up-zoning). In Köniz, so far, this instrument has not been used for affordable housing provision however |
| Pre-emption rights | – |

(continued)

**Table 5.5** (continued)

| | *Case of Köniz* |
|---|---|
| **(3) Contracts**<br>Long-term ground leases | The city of Köniz provides long-term ground leases on public land to non-profit housing associations. The provision of ground leases is connected to the fulfilling of social tasks e.g. affordability-, tenure security and social mix rules |
| Urban development contract | – |
| Tenancy matters | Besides the Swiss federal tenancy obligations, the municipality of Köniz has not introduced additional tenancy matters to better protect tenants from rent increase and dismissal |
| **(4) Property rights**<br>Purchase of land | Since many decades, the municipality of Köniz has been active in purchasing private property for the provision of affordable housing. Strategic acquisitions were made, for instance, in case of the 'Hertenbrünnen', 'Am Hof', or 'Dreispitz' settlements |
| | ***Case of Kloten*** |
| **(1) Supply-side subsidies**<br>For non-profit housing associations | The city of Kloten provides subsidies to non-profit housing cooperatives in the form of advantageous mortgages, direct loans and grants so that non-profit coops can afford e.g. ground leases on public land or the purchase of private land. Same as in the other three municipalities, the provision of subsidies is connected to the condition that these non-profit associations follow the cost-rent principle and social objectives such as social mix and occupancy rate rules |
| For private individuals | See city of Zurich |
| **(2) Zoning**<br>Special land use zones | In Kloten, the city does not use 'special land use plans' for affordable housing provision. Instead, the instrument is used to allow private investors to increase density outside the regular zoning plan. In any case, 'special land use plans' are generally coupled with the fulfilling of urban design and energy efficiency standards |

(continued)

**Table 5.5** (continued)

| | *Case of Kloten* |
|---|---|
| Quotas | The city of Kloten, so far, has not used a quota for affordable housing provision. However, in spring 2020, the voting majority will vote for a local referendum which aims to introduce a min. quota of 25% non-profit housing property by 2040. At the time of investigation, the voting has not yet taken place |
| Added land value capturing | See city of Zurich |
| Pre-emption rights | See city of Zurich |
| **(3) Contracts**<br>Long-term ground leases | The city of Kloten provides ground leases on public land to non-profit housing associations. The provision of ground leases is connected to the fulfilling of social tasks e.g. housing affordability and social mix rules |
| Urban development contract | – |
| Tenancy matters | See city of Zurich |
| **(4) Property rights**<br>Purchase of land | – |

the municipality initiated the founding of municipal foundations (e.g. "Stiftung Einfach Wohnen" in 2014). Thereby, the municipality seeks to raise awareness for social interests such as the introduction of social mix, income, and occupancy rate rules (Interviewee 40, City of Zurich, Urban Development Department, Expert in housing issues, July 31st 2019). The municipal authority has also intensified the use of "special land use zones" in recent years to provoke "room for negotiation" within building zones. In particular, to force private investors to promote social objectives. For instance, in relation to construction quality, urban design, and affordable housing goals if investors aim to benefit of a density increase outside the regular zoning plan (Interviewee 45, City of Zurich, Head of Planning Department, October 24th 2019).

The Case of Basel-City

To boost business and urban growth, so far, the city of Basel has primarily activated policy instruments which do not hamper private developers' interests to invest.

> So far, we have not had the same densification pressure as Zurich. We had the possibility to redevelop many unbuilt industrial zones. In fact, after the

> 80s and 90s - a period of structural decline - we have promoted population growth to promote employment increase in the city. (Interviewee 41, City of Basel, Head of Housing Department, August 20th 2019)

However, as tenants' social exclusion processes have increased in the last decade, Basel's local tenants association has initiated two local referendums to revise the Local Constitution in order to promote affordable housing (Interviewee 38, Head of Basel Tenants Association, June 26th 2019). Both initiatives aim to improve the living conditions for low-income and vulnerable groups such as old-aged and young families. Following the revised Constitution which was approved by 62% of the voting majority on June 10th 2018, the following changes must be incorporated in the Local Planning and Housing Act:

- Increase in the provision of supply-side subsidies to non-profit housing associations through the initiation of a municipal foundation for affordable housing.
- Introduction of a min. quota of 25% non-profit housing property by 2050 in the Local Constitution.
- Introduction of a more progressive land acquisition strategy for affordable housing.
- Stronger protection of tenants in case of rent increase after modernization through the introduction of rent levels which landowners must follow up to five years after the densification task is finished.

Through obtaining these measures, the city government promotes a more active land policy strategy in order to increase the share of affordable housing in the long run.

> In Basel-City, the political intention with the two constitutional initiatives is to introduce higher legal requirements for private homeowners for modernization and to hinder social exclusion of tenants that have lived in their dwellings for many years. (Interviewee 44, City of Basel, Planning Department, September 20th 2019)

In addition to the introduction of new instruments, the municipal planning authority uses available zoning measures (e.g. urban development contracts) in a more strategic way for effective affordable housing provision.

> In future terms, we will use urban development contracts more frequently to remain flexible and because we do not need a parliamentary decision to change something all the time. We do not want to dependent on the ideas of current political majorities in every project. (Interviewee 44, City of Basel, Planning Department, September 20th 2019)

The Case of Köniz

To promote affordable housing, on February 12th 2017, Köniz' voting majority agreed to revise the Local Zoning Act by adding two mandates. First, the city council must ensure that long-term ground leases on public land are provided to non-profit coops and that subsidized associations approve to the cost-rent principle. Second, on private plots larger than 4000m$^2$ floor area, the city can oblige the investor to provide min. 20–40% of the newly built apartments created through densification measures for affordable housing. Otherwise, a density increase outside the regular zoning plan is not being approved.

> For us, socially-sustainable densification does not only mean higher quantity but also higher social quality. (Interviewee 42, City of Köniz, Local Planner, August 20th 2019)

Moreover, to effectively steer local housing development, the city of Köniz has strategically purchased centrally located parcels.

> We perform an active land policy strategy. We discuss where the key parcels are to support urban development. [...] We purchase and sell land, but mainly we purchase. We do have a lot of public land reserves on which we can determine the use conditions. (Interviewee 42, City of Köniz, Local Planner, August 20th 2019)

In summary, Köniz' local planning authority combines public and private law instruments and is aware of how to use available policy instruments effectively to promote affordable housing.

The Case of Kloten

Same as Basel, so far, the city of Kloten has activated policy instruments which do not forcefully intervene into private investors investment interests (e.g. supply-side subsidies) in order to stay competitive and to attract business.

> In case we have public land, we collaborate with non-profit cooperatives. However, this is not the normal case. (Interviewee 29, City of Kloten, Head of Local Planning Department, August 2019)

Moreover, available zoning instruments such as "special land use zones" have been used without determining social restrictions for private investors to promote private investment. For example, in the "Waldeg-gweg" project, the landowner was allowed to triple the number of apartments on the same parcel without any obligation to fulfill social tasks e.g. related to affordable housing, the prevention of social exclusion or secure tenancy.

> With the initiation of special land use plans, we buy the right to participate and to have a say. [...] With this instrument, we can increase density and oblige the landowner to follow certain requirements related to architecture, urban design or energy goals. (Interviewee 29, City of Kloten, Head of Local Planning Department, August 2019)

The municipal planning authority legitimizes this strategy by the argument that the municipality seeks to attract business in order to compete with other suburban municipalities, and to become a regional center next to Zurich airport on its own.

> We support densification and modernization through raising incentives for landowners. For instance, investors are allowed to double or even triple the number of apartments on the same parcel. Thereby, we promote demolishment and rent increase of affordable apartments. [...] This procedure is politically and economically promoted by the local government. (Interviewee 29, City of Kloten, Head of Local Planning Department, August 2019)

In summary, Kloten municipality follows a land policy strategy for affordable housing which is indeed one-sided: under scarce land conditions and the parallel prediction of population growth (+50% by 2030), the rights of tenants are neglected while the power and wealth of the local growth coalition between the city government, private landowners and the local building industry increases even further.

### 5.4.5 *Four Legal Intervention Ways to Generate Affordable Housing Outcomes*

The main question introduced in this article addresses the link between planning and affordable housing provision as follows: *How do municipal planning authorities promote affordable housing in densifying cities?* In the previous chapter, we show how institutions in general, and the strategic use of specific land policy instruments in particular, are at core for answering this research question. Not only does the article reveal how an instrument's effect on affordable housing provision is indeed very different between the four municipalities. Moreover, our results show that the mere availability of land policy instruments is not sufficient for the effective provision of affordable housing but that the municipal planning authorities' strategic activation of specific instruments matters.

1. *Policy Instruments That Regulate Land Uses Using Public Policy with No Direct Impact on The Use Rights of Land (Supply-Side Subsidies)*
   Our analysis in four Swiss municipalities reveals that, so far, public policy instruments with no direct impact on the content of land use rights have proved to be the preferred support mechanism for the provision of affordable housing. This is because supply-side subsidies do not have a direct impact on the private property owner's freedom or investment interests. As a consequence, public subsidies are easier to implement for municipal authorities than tools which intend to change property rights. City governments do not need to gather the political majorities to oppose private development rights which is why the whole political spectrum (from left-wing to conservative parties) is more willing to agree on. As our analysis reveals, however, these rather weak instruments are not sufficient to provide affordable housing under scarce land conditions. They need to be supplemented by more interventionist land policy instruments to provide housing for all income groups (see next paragraphs).
2. *Policy Instruments Using Public Policy Leading to A Regulation of Use Rights on Formal Ownership (Zoning)*
   In the municipalities investigated (two core cities and two suburban municipalities), zoning mechanisms are acknowledged to be very effective in steering land use for affordable housing. This is because, when land gets scarce, planners are in need to actually intervene into private property owners rights to have a say how, for the

benefit of whom, and for what existing housing stocks should be (re)developed. For example, a popular zoning mechanism which seems to succeed in promoting affordability objectives effectively is the use of quotas. Although quotas do not lead to a direct intervention into private ownership, they help local authorities to communicate long-term planning goals and to legitimize the reinforcement of new planning measures. However, at the municipal level, there occur differences how such additional zoning instruments are strategically implemented. In Zurich, for example, where the political majority for more proactive ways of land policy exists, zoning instruments which provoke "room for maneuvering" on private property have more intensively been used in recent years. Specifically, the zoning instrument of special land use plans as it effectively intervenes into market forces. These zones grant municipal authorities the right to distribute, remove, and relocate private development rights according to social and affordable housing needs. In contrast, in the suburban municipality of Kloten, where political majorities follow a more liberal tradition of state intervention and the share of public ownership is low, our analysis shows that special land use zones are used in a different way. Here, "special land use plans" are used to promote modernization of existing housing stocks but at the expense of its social side (e.g. affordable housing) in order to attract business and to stay competitive.

3. *Policy Instruments Leading to A Legal Redefinition of Property Rights (Contracts)*
In the four municipalities analyzed, the use of policy instruments which lead to a legal redefinition of property rights have gained in strategic relevance in recent years: when urban land becomes scarce, the inertia of private landownership might be too strong which is why increased flexibility but also planning security and predictability is needed. As a consequence, all forms of "planning by contract" such as ground leases, urban development contracts, and changes in tenancy matters help planners to effectively integrate their visions and housing policy objectives into private development plans. Results show that especially in cities which do not have much public land reserves such as Kloten, this flexible type of planning has gained in importance. Through the possibility to strategically negotiate the terms and conditions, public–private-partnerships help planners to increase their power in front of landowners.

4. *Policy Instruments That Redistribute Property Rights (Public Ownership)*
   Landowning municipalities such as Zurich, Basel, or Köniz succeed in effectively promoting affordable housing because they benefit from the power granted by property rights. As landowners they are able to steer land use according to their socio-economic interests and visions. However, such active land policy strategy comes with a number of problems too (Gerber et al., 2017): *first*, a municipality needs to be able to finance such acquisition strategy, which is difficult to manage especially for smaller suburban cities such as Köniz or Kloten. For them, the challenge with high land prices and austerity imposed on public actors is even harder to handle than for core cities such as Zurich and Basel. Land deals might also be financially risky for the public sector. Therefore, the question arises whether municipalities should take these risks or better transfer them to the private sector. *Second*, the city government needs to convince the parliament and the population of the benefits that emerge through a public authority managing assets. This mission is especially difficult for more liberal municipalities in which the political spectrum is more likely to agree on less state intervention (e.g. in Kloten). Our results show, however, that once the city government agrees on a more proactive intervention strategy for affordable housing provision such as in Köniz, the spatial development opportunities that arise through it are quickly recognized by politicians and residents. In the city of Zurich, for example, the purchase of public land has enabled municipal authorities to provide affordable housing through various ways such as the construction of public housing or the provision of long-term ground leases to non-profit coops. As a result, the share of social housing constantly increases and the financial expenses for the support of social welfare recipients decreases simultaneously.

In summary, results show that Swiss municipalities do not follow a "one-solution-fits-it-all" land policy strategy for affordable housing. Depending on the socio-political context (e.g. district characteristics, financial capacity, political majorities, or the cultural conditions related to urban regeneration goals), planners follow heterogenous policy goals and try to promote housing affordability by implementing different policy instruments. However, we summarize that an active municipal land policy strategy for effective affordable housing provision requires both—the combination of public and private law instruments and the strategic activation of them: since private property rights are strongly protected by the Swiss Constitution and very inflexible, Swiss municipal land use planning seems to experience difficulty in implementing democratically accepted spatial development plans on titleholders due to conflicting interests. As a consequence, the real housing challenge is not so much plan making, but rather plan implementation. Without heavy state intervention such as expropriation, new housing regulation in favor of housing affordability (e.g. new zoning) only gets implemented when titleholders agree to undertake new developments, sell their stock or the land, or transfer their housing development rights.

The shift towards densification in land use planning makes this conflicting relationship between policy intervention and property rights even more difficult in the four cities investigated since densification implies to deal with the already built environment. Planning therefore takes place within a tight web of existing rights and duties engraved in complex institutional norms and regulations. Potential for redevelopment is often given, but the land is frequently not accessible due to the land rights secured by strongly protected property titles. Under these circumstances, planners often fail to deal with complex private property rights arrangements as most public intervention ways were crafted to handle simpler property rights situations on unbuilt agricultural or industrial land. Therefore, to cope with complex property rights situations on already built land such as intermixed parcels of different sizes, co-ownership constellations, rights to object granted to neighbors, rights of way, or mosaic of easements, more than ever, planners need a keen understanding of the close interactions between public policy and property rights to effectively steer affordable housing development. Our analysis reveals that it needs all the finesse and competencies (e.g. knowledge, financial resources, networks, personnel) of municipal planning administrations to implement affordable housing objectives in dense cities, because landowners have the power to defend the status quo through veto rights.

In core cities like Zurich, for example, public officials succeed in increasing the share of affordable housing units as they rely not only on zoning but also municipal ownership, long-term ground leases, and tenancy law. In smaller municipalities such as Kloten, however, expert knowledge as well as personnel and financial resources for strategic activation of these policy instruments is not as pronounced. Also, the political acceptance and majorities for more proactive forms of planning is not always given—especially in more suburban areas. Local politicians often regard offending private investors' plans as a too risky business for the municipality's financial situation. This leads to the conclusion that even though no general local intervention strategy for affordable housing provision exists, this study has indicated how municipalities might coordinate and strategically activate different policy instruments to deal with the scarcity of land and to satisfy affordable housing needs more effectively in the long run. Indeed, we showed that the introduction of new policy instruments is not always necessary but that the strategic activation and combination of available instruments is becoming more relevant.

### 5.4.6 *Towards an Active Municipal Land Policy* with *Property Rights*

While there is a growing body of literature focusing on the social impacts of densification on households (Bramley et al., 2009; Burton, 2000, 2003; Chiu, 2003), and another extensive body of research looking at land policy issues for the management of natural resources (Gerber et al., 2018; Hartmann & Spit, 2015; Ostrom, 2007), research on how to combine the two concepts to housing inquiries is still thin (Balmer & Gerber, 2017; Nicol & Knoepfel, 2008). More qualitative and quantitative research is still needed on whether or not specific policy instruments such as changes in tenancy matters or the property rights logic can effectively steer affordable housing, and why some municipal authorities decide to activate specific instruments while others do not.

In this article, we introduced a neoinstitutional analysis framework which postulates a causal relationship between (1) the affordable condition of the housing resource, (2) the institutions in force and corresponding policy instruments, and (3) the involved actors and their appropriation strategies. We analyzed the mechanisms at play between these three variables that explain why some groups or interests experience disproportionate access to the decision-making process on housing use and tend to lose while others win. An active land policy strategy which aims to promote affordable housing through the activation of both new and available instruments proofed to be become particularly relevant in this matter. In particular, we showed how different land policy instruments function and are strategically activated by municipal planning authorities to provide affordable housing. Even though our results are limited to four Swiss cities, potential for generalization results from the following identified causal mechanisms which are expected to have broader significance in other urban contexts too: affordable housing provision results from the intertwined relationship between land use planning (public policy) and property rights—the two main sources of formal constraints. Planners can influence the private property owners' behavior in favor of increased housing affordability if they are able to find ways which reinforce their position in front of powerful landowners. To do so, they need to activate public and private law instruments which do not always need to limit property owners' rights but also work *with* property rights.

This study addresses a gap in housing study literature (Aalbers, 2017; Bramley et al., 2009; Burton, 2000, 2003; Chiu, 2004; Kadi & Ronald, 2014) as it analyzes the cities lack of affordable housing as a land policy issue, and in relation to the formal institutions and the municipal authorities' decision-making strategies involved. Taking into account future challenges of land scarcity that currently evolve in many cities (Gennaio et al., 2009; Touati-Morel, 2015), the findings of this study may help municipalities to counteract trends of rising commodification and financialization of urban housing stocks. If city authorities do not succeed in providing affordable housing in densifying cities, the preservation of urban social qualities such as social mixing, tenure security, or community cohesion is in acute danger since more and more tenants are forced to leave due to rising rents after densification. This scenario is highly unsustainable. This article helps municipal planners, practitioners, and policymakers to prepare for future housing challenges: a stable "right-to-housing" for all does not necessarily require the mere introduction of new policy instruments but the strategic activation of available instruments matters.

## 5.5 Article 4: The Commodification of Temporary Housing in Zurich

Gabriela Debrunner & Jean-David Gerber
Journal: *Cities*, Vol. 108, https://doi.org/10.1016/j.cities.2020.102998
Impact Factor (2019): 4.802

Status: Published

**Abstract:** Since the 1970s, temporary uses of vacant spaces have become a preferred urban development strategy to revitalize centrally located neighborhoods. In the housing sector, however, temporary uses are barely registered as they provide only short-term shelter in buildings shortly before demolition. Therefore, they do not secure a stable right to housing. In Switzerland, nevertheless, temporary uses are increasingly gaining momentum in the housing segment. Since the 2010s, besides institutionalized but non-profit temporary housing, a for-profit model has emerged. This commodified model is managed on the owners' behalf and is based on loaning law contracts that require payment for operating costs, but not rent. Consequently, the legal protection of the temporary users' rights, namely low-income families, single parents, people with social aid, and students remains weak. This article detects the mechanisms at play explaining the reasons for the shift towards profit-seeking in temporary housing by using an institutionalist and actor-centered analysis approach. Through a qualitative single case study analysis of Zurich, Switzerland, the phenomenon will be analyzed in a city confronted with increasing affordable housing shortage and densification pressure.

*Keywords: Temporary use; housing affordability; housing policy; commodification; urban social movements; densification*

**Research Highlights:**
*The paper…*

- *…examines an extreme case of affordable housing scarcity in a context of densification – the emergence of a for-profit temporary housing model in the city of Zurich, Switzerland.*
- *… explains in detail why and how the actors involved – in particular, landowners, public authorities, mediators, NGOs, and residents – benefit from the emergence of such new housing model.*

- *...shows that not only young and flexible people live in temporary housing locations shortly before demolition, but also young families, single parents, and low-income migrants.*
- *...questions the role of property owners, mediators, and the city government in promoting for-profit temporary housing and explains how they are able to defend their interests.*

### 5.5.1 Temporary Housing in Densifying Cities

Since the 1970s, temporary uses of vacant spaces have become a preferred urban development strategy to preserve and to revitalize centrally located neighborhoods, to express political claims, and to boost economic and social innovation in cities (Castells, 1983; Florida, 2002; Galdini, 2019; Oswalt et al., 2003). Whether used for illegal or publicly subsidized temporary use, urban scholars have recognized the significant role of temporary urbanism for the dynamic (re)production, transformation, and distribution of space in order to support self-determination, diversity, and flexibility of today's urban society (Amin & Thrift, 2002; Bishop & Williams, 2012; Colomb, 2012; Smith, 2017).

Temporary use, however, is neither clearly defined in its form of activity nor in its duration or legal dimensions. The only common characteristic is temporariness, which means that temporary uses are "explicitly" and "intentionally" time limited in nature (Lara-Hernandez et al., 2020: 1; Németh & Langhorst, 2014: 144). Unlike short-term rentals or Airbnb-arrangements (van Holm, 2020), "temporary housing" as defined in this study takes place in vacant buildings shortly before demolition or reconstruction. It refers to an undefined temporary gap between the former residents' moving out and the demolition and/or renovation of the building. Unexpected events such as the delayed approval of the building permit or changing investment conditions may lead to an expansion of the temporary housing period which are not previously foreseen (Angst et al., 2008).

In the Swiss context, which will be presented in greater detail in the following sections, we observe the situation that temporary solutions have increasingly gained momentum in the housing sector. For low-income residents, temporary housing offers the possibility of living centrally and at low cost. Particularly in cities, we observe that the dynamics of temporary housing have changed since the revision of the Federal Spatial

Planning Act (Art. 1, SPA) in 2013. Following the revised Act, Swiss municipalities have become obliged to promote inward settlement development to curb urban sprawl. Through a process of densification (also termed "intensification" or "consolidation") leading to an increase in the number of households within existing municipal boundaries, urban land gets increasingly scarce and the competition to use this land is rising. Confronted with such tight market conditions, we identify a housing situation in which a new, profit-oriented temporary housing model that is managed on the owners' behalf has emerged. As our analysis reveals, this new model of temporary housing is based on loaning contracts that require payment for operating costs, but not rent. Despite this legal shift leading to the erosion of the protection of tenants' rights in the name of increased flexibility for landowners and developers, this new business model proves to be favored by a coalition of all major actors involved—temporary users, property owners, mediators, and municipal authorities. This article interrogates this puzzling situation and aims to explain the reasons behind.

To capture this phenomenon, we ask: *What are the institutional arrangements making for-profit temporary housing possible? Which rationales of the different actors involved in this system explain its expansion? And what are the consequences on the different categories of actors?* We identify the mechanisms at play explaining how the shift towards profit orientation in temporary housing took place and discuss how the results are to be interpreted from a critical urban development and social justice perspective. To answer the research questions, we apply an institutionalist and actor-centered analysis approach and qualitative single case study methodology.

### 5.5.2 Development Stages of Temporary Housing

To explain the emergence of a for-profit temporary housing model, the article applies an institutionalist and actor-centered analytical approach (Healey, 2007a; Nicol & Knoepfel, 2008): following this approach, housing is regarded as a resource, the affordable status of which is seen as the result of a complex interplay between the local regulatory framework and the decision-making behavior of the actors involved (municipal authorities, landowners, etc.). We distinguish two main sources of institutional rules: public policies and property rights. Institutions are defined as a set of norms and values—formalized in legal rules or not— that structure humans' expectations about what others will do (Hall & Taylor, 1996: 956). Within an institutional setting, actors develop strategies to defend their own interests in order to meet a particular goal (e.g. affordable housing provision) (Gerber et al., 2018).

More precisely, through public policies (stipulated in public law), public actors are granted democratic legitimacy and material power to solve a public problem in the name of a public interest, thereby confronting existing property rights (stipulated in private law), which follow an opposite logic—the protection of private interests against the state (Knoepfel et al., 2011). For example, municipal authorities provide the local policy framework regulating temporary housing, e.g. through zoning or the authorization to use the premises, as well as through specific measures designed to promote affordability. However, property owners are in a position to decide on temporary uses of their site. Their property titles, which grant them the power to define use or transfer rights, can oppose the interests defended by public actors. Hence, even though potential for affordable housing provisions is often given, apartments are often not accessible due to powerful landowners' interests (Gerber et al., 2018).

While temporary housing organized through public support has been known for several years (e.g., in the form of temporary student or refugee accommodations), a new trend towards commodification of temporary housing is gaining momentum. We will explain this shift by analyzing the underlying institutional rules as well as the rationales of the actors involved.

*Legal Security Through Institutionalization*

Historically, temporary housing has its roots in illegal and informal squatting. During the 1980s in many Western European cities such as Berlin, Amsterdam, or Brussels, illegal squatting of vacant residential buildings became an eloquent symbol for social protests against the scarcity of affordable urban housing and rising unemployment (Holm & Kuhn, 2011; Pruijt, 2013). Because of the failure of city councils to respond to the lack of affordable housing, non-profit grassroots organizations were founded out of the illegal squatting scene to provide an alternative to market-oriented housing (De Decker, 2009).

In the following decade, however, many urban governments started to support these grassroots organizations and to publicly finance alternatives to squatting themselves. They aimed to calm down the protests and decided to integrate these self-help initiatives into their social housing policy strategies. In other words, out of the informal, urban squatting movement an increasing number of activists' groups were clearly channeled into more stable and formalized patterns to keep a clean, safe, and respectable image of the city (Martínez, 2013; Mayer, 2007; Özdemirli, 2014; Priemus, 2011; Pruijt, 2003).

The first legal temporary housing agencies working on a non-profit basis were founded in this context. These mediator agencies are acknowledged as a primary institutionalized form of temporary housing in the sense that they started to connect vacancies and potential temporary users with each other on a professional basis. They also began to provide knowledge on the local legal framework for political, organisation, technical, administrative, and contractual constraints. Contrary to illegal squatting, they benefited from legal security, stability, and financial support as they operated under legal norms and political-administrative procedures. The city council used its authoritative power, including regulatory statues, penal law, or social housing policy regulations to promote non-profit temporary housing. This procedure served as a favorable public coping strategy to deal with the urban affordable housing shortages during the 1980s and 90s in many Western European states (Colomb, 2012).

In Belgium, for instance, so called "social rental agencies" have started to integrate temporary housing as a widespread means to support vulnerable households covering a large part of the affordable housing demand in the country during the 1990s and 2000s (De Decker, 2009). In the Netherlands, "anti-squat" organizations have been created to temporarily manage buildings and to protect them from vandalism (Priemus, 2011; Pruijt, 2013). In Eastern Germany, the model of "safeguard houses" (in German: "Wächterhäuser") was successfully implemented at the beginning of the 2000s to offer affordable housing options to temporary users who, in return, protect the abandoned properties from decay (Dubeaux & Sabot, 2018). In summary, these temporary housing agencies all work on a non-profit basis to meet the needs of specific categories of tenants. As social organizations, they are integrated into the municipal housing policy system and not only provide accommodation to low-income groups (students, young adults, social welfare recipients, etc.), in most cases, they also help them to manage their daily life in a broader sense (budgeting, household organization, etc.).

*Economic Profitability Through Commodification*

As described in the previous section, the integration of non-profit temporary housing into urban housing policy systems has become a favored strategy to deal with affordable housing shortages and short-term vacancies for the past three decades (Cardullo et al., 2018; Vallance et al., 2017). Simultaneously, urban housing stocks have become the main target of capital investment and a safe source of revenue, especially for landowners. Such commodification strategies, however, potentially hamper affordable housing provision (Aalbers, 2017; Harvey, 2012; Marcuse, 1985): the concept of "commodification" is a very old one and acquired its meaning with the writing of Karl Marx (1859). It describes the process of how housing is influenced by market and profitability objectives which not only determine what type of housing is built but also how it is used, managed, and distributed (Harloe, 1982: 40). Commodification of housing relies on the assumption that the market, including the profit-maximizing rationality of investors, is the most efficient solution to guarantee the provision of housing for all income segments (Kadi & Ronald, 2014; Rolnik, 2013). Housing is no longer considered a basic human need and essential good, but rather more a commodity that must be traded or paid for in a globalized financial market (Harvey, 2005).

Critics point out that commodification objectives in the housing sector have profoundly affected the acknowledgment of the "right to housing" (Brenner et al., 2012; Harvey, 2012). While the process might be advantageous for those select few who reap the disproportionate benefits of the capital gain, the vast majority—and particularly those of lower income—would have little ability to capture value from this process (Harvey, 2005: 166). Moreover, the investors' profit-seeking behavior and urban (re)development objectives foster gentrification processes as they lead to higher rents and to risks of social exclusion, tenure insecurity, and segregation in cities (Chiu, 2004; Korthals Altes, 2016; Lees, 2008). The role played by private landowners becomes particularly relevant in this matter. Due to the protection guaranteed by private property titles, landowners are free to define the profit margin to be targeted on their parcels and to set the rents according to market prices (Marcuse, 1985). Real estate property is hence acknowledged as a foundational element to both power and wealth (Aalbers & Christophers, 2014) as it shapes the level of social inequality and exclusion in cities (Galdini, 2019; Lai et al., 2018).

Commodification processes often lead to counter-strategies that aim towards decommodification. The concept of "decommodification" stands for the strength of social entitlements and for the citizens' degree of immunization from market dependencies (Kadi & Ronald, 2014: 270). It also aims to overcome the incapacity of generalized commodities to meet basic human needs for all as the process aims to reduce the control of the market laws on goods and thereby diminishes the pressure to generate financial profit (Lees, 2008). Consequently, decommodification of housing stands for a move away from the value of housing considered by its financial value to a focus on its use value in order "to provide every person with housing that is affordable, adequate in size and of decent quality, secure in tenure, and located in a supportive neighborhood of choice, with recognition of the special housing problems confronting oppressed groups" (Achtenberg & Marcuse, 1986: 476). In temporary housing, the shift towards profit-orientation has not taken place. There are only a few for-profit temporary housing models and detailed analyses are missing, although this housing type seems to be expanding. In the Netherlands, for example, for-profit private "anti-squat-agencies" manage buildings on the owners' behalf and enter contracts with residents following an "anti-squatter-attitude" (Priemus, 2011). The city of London is familiar with so called "safeguard houses", meaning that a private agency organizes temporary housing in vacant buildings before demolition (liveinguardians.com, 2020). It is unclear, however, why the involved actors participate in for-profit temporary housing, to what extent they only follow capitalist motives, and how these models have evolved. In addition, our study reveals that temporary users of for-profit temporary housing have a different profile and follow principles other than those of commercial or creative temporary users (e.g., such as artists, bohemians or start-up groups). They neither follow interests of creating an alternative lifestyle nor do they identify as members of the creative scene (Blumner, 2006). Because this turn towards profit-seeking in temporary housing is under researched, this article aims to analyze the institutional mechanisms and actors' rationales driving this shift.

To fill this gap, we focus the rest of the article on the Swiss housing situation. Switzerland makes an interesting study for the analysis of temporary uses in housing since the pressure on land and housing markets has increased in recent years, especially in cities. Swiss cities are characterized by extremely strained relations (vacancy rates below 1%) and rising rents (Balmer & Gerber, 2017: 8). People with lower incomes as well as the middle class are negatively affected by inadequate affordable housing supply (FOSI & FOH, 2015). As a consequence, temporary housing solutions as a flexible model to handle population and economic growth have gained momentum in recent years (Bürgin, 2017). The state is organized on three executive levels—the confederation, the cantons, and the municipalities. The municipalities hold the greatest decision-making power regarding spatial development: they grant building permits according to the stipulations of zoning regulations, which is binding to private landowners (Gerber et al., 2017: 1690).

### 5.5.3 *Temporary Housing in the City of Zurich*

This study relies on a qualitative research approach. Through an embedded single case study analysis, the mechanisms at play explaining the emergence of a for-profit temporary housing model will be assessed. This approach makes a detailed and differentiated understanding of the studied phenomenon possible (Yin, 2018: 15).

*Case Selection*
The city of Zurich is a German-speaking city and the largest urban center in Switzerland (Statistics City of Zurich, 2020). The municipality of Zurich represents both the core center of Zurich agglomeration and the capital of the canton of Zurich. Zurich makes an interesting case study for the analysis of temporary housing as temporary uses of all kinds (e.g., commercial, office, or residential) have increased in recent years (Bürgin, 2017). In 2016/2017, Zurich was identified as a hub of temporary uses within Switzerland. In total, 417 temporary use projects were set up in Zurich[16] whereas in the same period, only 61 temporary uses were registered in Basel, 37 in Berne, and 6 in Geneva (Wüest & Partner, 2017).

This predominance of Zurich in the temporary use scene is connected to the city's steady population and economic growth, but also to its function as an international investment center (Theurillat & Crevoisier, 2013).

[16] The study mentioned does not distinguish between housing or commercial temporary uses.

Since 1980, Zurich's population has increased by +17% and investment into real estate has constantly risen (Statistics City of Zurich, 2020). During the 1980s, Zurich has started to deindustrialize and some of the abandoned industrial land was redeveloped or taken over for cultural and temporary activities. Many investors and developers started to recognize the city's economic potential and began to reclaim the buildings they owned (Rérat & Lees, 2011: 131).

Since the year 2000, however, the (re)development of existing buildings or vacant plots has become increasingly challenging for investors since inner-city greenfield and brownfield sites are missing. Only 10% of all newly built apartments have been built on unbuilt parcels during the last two decades. Most of the newly built dwellings have been created through reconstruction and densification of existing housing stocks on already built land (Statistics City of Zurich, 2020). As a consequence, the tensions between densification, modernization, and social exclusion mechanisms have increased significantly in recent years since redevelopment initiatives have led to higher rents and new (temporary) housing forms (Rérat, 2012). In fact, the absolute number of social evictions due to redevelopment in the city's private rental sector doubled within the period of 2006–2017 (Statistics City of Zurich, 2017). Between 2000 and 2013, rental prices in the housing stock increased by 37%, while rental prices on the free market rose by 75% (Balmer & Gerber, 2017: 8). Hence, moving to cheaper suburban areas or new forms of temporary housing have remained the only option for many vulnerable and lower-income groups in Zurich.

*Case Study: Historical Background of Temporary Housing in Zurich*
Temporary housing has a long tradition in Zurich. This housing type first appeared during the "1980-opera-riots", an urban social movement triggered by the tense situation on the housing market and unfulfilled expectations about urban cultural life and open spaces (Kriesi et al., 1995). On 30 May 1980, more than 200 young people (most of them under the age of 25 years), demonstrated against unequal policy investments in front of the Zurich opera house. The protest ended in a riot between the police and activists, and a two year political struggle about the support of alternative living forms and cultural activities. As a consequence, temporary squats of vacant houses became an eloquent symbol for the youths' protest against Zurich's Fordist model of economic growth, the lack of urban affordable housing, and rising unemployment.

Aiming for progressive political change, Zurich's youth squatted buildings in order to protest against the predominance of economic interests, the growing gentrification of inner-city neighborhoods, and the privatization of urban housing stocks (Holm & Kuhn, 2011).

The city government, however, publicly criticized the political attitude of the growing squatting scene. In most cases, it did not tolerate the status of illegality in housing and used police force and violence against illegal housing squats (Stahel, 2006). The civil society reacted strongly against this procedure and forced the political elites to undertake policy changes, particularly regarding the expansion of social housing as well as the introduction of new security policies. In the following section, we will explain how these policy changes during the 1980s and 1990s still affect Zurich's temporary housing practice today.

*Methods*

We conducted multiple methods to understand the mechanisms at play explaining the emergence of for-profit temporary housing in Zurich. The methods used enabled us to grasp a largely unknown and still barely quantifiable phenomenon (George & Bennett, 2005). The empirical data was collected in three steps (Table 5.6).

As a *first step*, we analyzed the public regulatory response to Zurich's temporary housing situation. We strived to explain what public policy interventions (e.g., planning, housing, security, and social welfare policies) are involved in temporary housing as well as what private law institutions (e.g., property rights, tenancy matters) guide and shape the emergence of this housing type. We started with the analysis of socio-economic statistical data to explain the city's housing situation over time (development of rents, vacancy rates, housing prices). In addition, we analyzed the local institutional rules involved in regulating temporary housing through an in-depth qualitative analysis of policy documents (parliamentary debates, legislations, government reports). These methods were applied to capture the institutional origin and functioning of temporary housing within its real-life socio-economic and -political context (George & Bennett, 2005).

In our *second step*, we aimed to understand the objectives and strategies of the actors involved in temporary housing (public authorities, property owners, mediators, temporary users) to explain the emergence of for-profit temporary housing from an actor-centered perspective. Therefore, within the timeframe of January 2015 to June 2019, we conducted 25 semi-structured interviews with experts and residents to understand their

**Table 5.6** Steps, aims, and methods employed in this study

| | *Research questions* | *Research aims & actions* | *Methods employed* |
|---|---|---|---|
| Step 1 | RQ1: What are the institutional arrangements making for-profit temporary housing possible? | Analyse public regulatory response to Zurich's socio-economic and housing situation, with a special focus on the interactions between regulation and property relationships | Analysis of statistical data, policy documents (parliamentary debates, legislations, government reports), newspaper articles, and 'grey' literature |
| Step 2 | RQ2: Which rationales of the different actors involved in this system explain its expansion? | Analyse actors' strategies related to temporary housing over time, with a specific focus on temporary users (residents) | 25 semi-structured interviews with experts as well as with temporary users |
| Step 3 | RQ3: What are the consequences on the different categories of actors? | Analyse the socio-economic housing situation of temporary users and reflect on the repercussions of Zurich's housing policy strategy for the stakeholders involved | Unstructured field observation of temporary housing sites, questionnaire with temporary users including evaluation of income and paid rents |

interests and perspectives. We performed interviews to gain information about "how" and "why" these actors defend their goals in temporary housing to get to know their motives (Yin, 2018: 118). More precisely, we conducted interviews with seven temporary users, six property owners, six mediator agencies (non- and for-profit), and one representative of the city department for housing. In addition, we interviewed one expert from the local tenants' association, one from the local homeowners' association, one expert from a private local real-estate management agency as well as two politicians of the local legislative parliament. All interviewees were chosen due to their detailed understanding and knowledge of the topic as well as based on their practical expertise related to the position they occupied within certain professional structures (Yin, 2018: 118). For instance, the actors representing the temporary users, the property owners, and the mediator agencies were chosen as they were part of seven ongoing temporary housing projects in Zurich city region (Table 5.7). In doing so, we aimed to gain knowledge from participants directly involved in—or affected by—temporary housing. We stopped interviewing people when no new insights from data gathering were collected because the answers of the respondents coincided (Yin, 2018: 118). In a *third step*, we focused on the temporary users' socio-economic housing situation to draw conclusions on the beneficiaries of for-profit temporary housing strategies. To do so, we carried out "unstructured field observations" (Althaus et al., 2009: 24) of the seven temporary housing sites to systematically document impressions related to the quality, size, and location of home and to provide material for the formulation of detailed questions during interviews. We noted all our observations in a field book to constantly improve our knowledge and to tailor questions prior to new interviews (Yin, 2018: 132). Moreover, we conducted a questionnaire with the temporary users, including specific questions capturing their socio-economic profile (age, gender, education/employment, income in relation to rent, household size, duration). The data collected makes it possible to compare the users' situation with each other (Table 5.7) and to reflect on potential repercussions for municipal policymaking. All temporary users interviewed signed a document for ethical approval to ensure that the data collected in their home can—in an anonymous way—be used for publication.

### 5.5.4 *The Commodification of Temporary Housing in Zurich*

The emergence of a for-profit temporary housing model in Zurich took place in two stages (Fig. 5.4). *First*, through a process of institutionalization, which demarcated it from squatting, temporary housing was no longer considered illegal. Temporary housing became regulated through formal rules and procedures, but with non-profit objectives. Actors involved in institutionalized temporary housing benefit not only from higher legal protection, but also from the ability to sanction abuses.

At the beginning of the 2010s, out of institutionalized non-profit temporary housing, *an additional step* took place towards the emergence of a commodified model. This step was connected to the legal obligation to densify within municipal boundaries in Swiss cities which has led to increasingly tight urban housing markets. Under scarce land conditions legal security becomes relevant for landowners, as well as planning predictability, flexibility, and economic security. To cope with building delays, temporary housing offers them a leeway to bypass tenant protection in housing and to realize profitable housing projects at central locations without substantial investment risks. In this model, private mediator agencies provide their services and knowledge to their clients

**Table 5.7** Characteristics of the seven investigated temporary housing projects in Zurich in the year 2015

| *Project Nr.* | *Temporary users' age/ gender* | *Education/ employment situation* | *Monthly net-income (CHF)/person* | *Household size* | *Type of mediator agency* | *Location* | *Duration* |
|---|---|---|---|---|---|---|---|
| 1. | 25 years, female | Student | 1000–1500 CHF | Shared flat with two temporary users | Non-profit | City Center | 2 years |
| 2. | 26 years, male | Student | 2000–2500 CHF | Shared flat with three temporary users | Non-profit | City Center | 4 years |
| 3. | 32 years, male | Professional | >3500 CHF | Shared flat with three temporary users | Non-profit | City Center | 6 years |
| 4. | 42 years, female | Social-welfare recipient | 2000–2500 CHF | Family household; single mother with two children | Non-profit | City Center | 1 year |
| 5. | 32 years, male | Professional | 1500–2000 CHF | Family household; couple with one child | For-profit | Urdorf | 10 months |
| 6. | 25 years, male | Student | 1500–2000 CHF | Shared flat with three temporary users | For-profit | Winterthur | 4 months |
| 7. | 27 years, male | Professional | 1000–1500 CHF | Single household | For-profit | Küsnacht | 6 months |

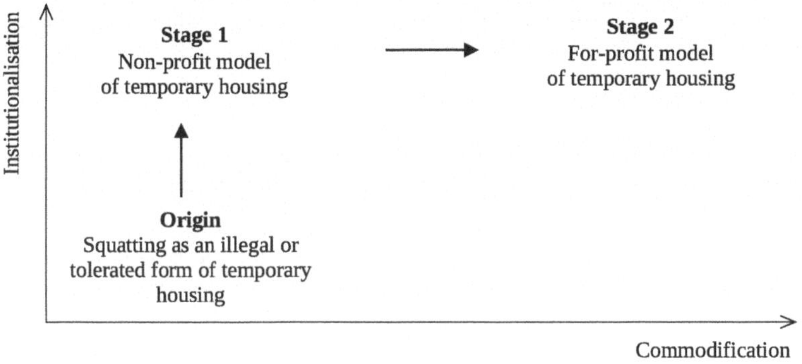


**Fig. 5.4** The emergence of for-profit temporary housing in Zurich can be explained by the twofold process of institutionalization and commodification

(property owners) with profit-oriented motives and get paid for their mediation work.

In the following section, we explain in detail these two mechanisms at play that lead to the shift towards profit-seeking in temporary housing by analyzing the local regulatory framework and the involved actors' rationale.

*Local Regulatory Framework*

In this section, we analyze the local regulatory framework (stressing both public policies and property rights) regulating temporary housing in Zurich. Temporary housing is addressed in several public law areas. We start with the city's police power. Then, we emphasize aspects of housing and planning policy because of their significant impact on housing (re)development. Finally, the role of private law will also be addressed as it has an impact on the property owner's decisions regarding the use of urban vacancies (Table 5.8).

Public Policies Regulating Temporary Housing in Zurich

The "1980-opera-riots" have marked a significant change in Zurich's police practice, particularly in regard to the clearance of housing squats. Following the claims of the activist groups, in the year 1989, the municipal parliament agreed to introduce a legally binding temporary housing clause in police law in order to calm down the youth protests of that time (Interview 20, member of municipal legislative parliament, 06.05.2015). Following the new legislation, the city police were only allowed to clear

**Table 5.8** Main dimensions of the regulatory framework influencing temporary housing practice in Zurich

| | *Regulatory measure* | *Content* |
|---|---|---|
| Police and security policy | 1989: Municipal Housing Squat Regulation | Police are not allowed to clear illegal housing squats without property owners' approved building permit, with the aim to raise political tolerance and awareness of alternative movements |
| Housing policy | 1924: Municipal Housing Policy Act, Art. 1–6<br>2011: Revision of Municipal Constitution, Art. 2 | New objective to increase the share of non-profit housing property to counteract rising rents and tenants' social exclusion mechanisms |
| Planning policy | 2018: Revision of Municipal Planning Act | Push for density increase within municipal boundaries to combat urban sprawl and resource depletion |
| Tenancy law | 1911: Federal Obligations Code, Art. 253–274 | To protect tenants against unfair dismissal and rent increase |
| Loaning law | 1911: Federal Obligations Code, Art. 305–311<br>2013: in Zurich first applied for temporary housing | To allow flexible use of vacant sites at low-cost without being obliged to follow the rules of tenancy (e.g., three month contract termination deadline, rent deposit payment, etc.) |

housing squats if the property owners could provide an approved building permit (Municipal Housing Squat Regulation[17]). Otherwise, the owners had to pay for the evacuation costs themselves (Interview 34, CEO of a for-profit mediator agency, 26.04.2019). In the following years, this legal change has led to increased public and political openness towards squatting and temporary use in Zurich (Stahel, 2006).

In housing policy, the opera riot led to several changes too. During the 1990s and 2000s, public subsidies for non-profit housing organizations increased with the primary aim to support vulnerable groups who struggle to find adequate housing (Municipal Housing Policy Act,[18] Art. 1–6). Many of these social housing organizations are still tightly connected

[17] See "Zurich Housing Squat Regulation of 1989" which is based on the Cantonal Police Act of 10 March 1831 (CPA 550.1).

[18] Municipal Housing Policy Act of July 9th 1924 (LHPA 841.110).

to the municipality and considered a political response to the demands formulated during the "1980-opera-riot". Furthermore, on November 27, 2011, following a popular municipal initiative, 75.9% of Zurich's voting majority agreed to raise the share of social housing property (public and non-profit cooperative) to at least a third (33.3%) of the total housing stock by 2050 (Municipal Constitution,[19] Art. 2). In 2019, the property share of non-profit housing amounted to 24.6% (City of Zurich, 2019a). To meet this policy objective, the city of Zurich builds on an active land acquisition strategy in favor of public housing and non-profit housing cooperatives (Balmer & Gerber, 2017).

Temporary housing, moreover, is influenced by planning law. As already mentioned in the introductory section, in March 2013, the Swiss voting majority approved the revision of the Federal Spatial Planning Act (SPA) which, among other purposes, aims to increase density within municipal boundaries. Following the revised SPA, the city of Zurich has updated its planning legislation in 2018. Particularly, the city council initiated planning measures such as the introduction of densification zones[20] to effectively promote population growth through internal settlement development and the efficient use of energy (Zurich City Council, 2013: 5). As a consequence, urban densification projects leading to redevelopment of existing stocks have intensified in the last five years (Interview 35, City Department for Housing, 03.05.2019). Under these circumstances, temporary uses of vacant residential properties shortly before demolition or reconstruction have become a favored coping strategy for landowners. Temporary uses help them to flexibly deal with short-term vacancies and potential planning delays in this context of densification (Bürgin, 2017). Simultaneously, however, residents' social resistance strategies and NIMBY-syndromes against densification projects have increased too as modernization often leads to higher rents and to social exclusion processes of lower-income groups (COSD Zurich, 2014).

#### Private Law Configurations: Institutional Differences Between Letting Vs. Loaning

The temporary use of housing stocks is not only influenced by public regulations. Private law also impacts urban housing development. Swiss

[19] Municipal Constitution of April 26th 1970 (LC 101.100).

[20] In these zones, landowners can realize higher exploitation rates (number of apartments) in comparison to the former Municipal Zoning Act (1999).

landowners are in a power position to decide on the use and disposal of their housing premises due to the constitutional guarantee of property (Art. 26, SC). Property rights can only be restricted if strict conditions are met. In any case, full compensation is to be paid (Art. 5, 26, 36 SC).

In contrast, the rights of temporary users are protected by the articles for tenancy matters in the Swiss Constitution (Art. 109 SC) as well as in the Federal Obligations Code of 1911 (Art. 253274 OC). Under the rules of Swiss tenancy, private landlords, temporary users, and mediators agree to sign a terminable rent contract. What we observe in temporary housing practice is that the mediators are renting the whole building from the property owner with a regular rental contract that is time limited. The mediators sublet the individual apartments to the final temporary users (sub-tenants). Through this institutional set up, the landlord has the guarantee to have his facilities empty at the date initially planned because the options granted by the law to object to any decision of the landlord are weakened due to the limitations of the subletting contract. Theoretically, the mediators as legal tenants are still able to have recourse against the owner in the cantonal tenancy court but in practice they do not because they need to maintain a good relationship with the owners for further housing options.

Strategic weakening of the tenants' position takes place above all through another institutional mechanism. In Zurich, besides tenancy law, loaning law regulation is also accepted to regulate temporary uses in the housing sector (Federal Obligations Code, Art. 305–311). Historically, loaning law was introduced in 1911 as part of the Federal Obligations Code to arrange temporary uses of all kinds in vacant places (e.g., in garages or tool rooms). In contrast to short-term rent or lease, loaning regulation allows flexible use of vacant sites at low cost without being obliged to follow the rules of tenancy (e.g., three month contract termination deadline, rent deposit payment, etc.). This also means that the user—legally the "borrower"—does not have to pay a fixed rent but rather a fee for monthly maintenance and operation costs such as water, heating, and electricity. The owner—legally the "lender"—has no restriction to keep the three month contract termination deadline, as is mandatory in Swiss tenancy law. This legal framing gives property owners the ability to force users to leave the property at any moment and at short notice as they do not have a legal standing to claim their rights in court. Additionally, lenders have no duty to provide maintenance, such as insulation, heating, or covering any damages in the apartment as would be mandatory under tenancy law. It is therefore possible that lenders use this legal discretion to evict borrowers on short notice. The temporary user, in addition, does not have the right to appeal against contract termination, unfair treatment, or other abuses in the cantonal arbitration board as legally, they are not identified as "tenants" (Büchi & Gehrig, 2014). In Zurich, loaning law was first used for the regulation of temporary housing projects in 2013. In contrast to other Swiss cities, there exists no legal restriction which forbids the use of loaning law for housing premises (see discussion for details).

Ironically, in practice, our case study reveals that temporary users develop resistance strategies against loaning practice since they do not have to pay a rent but instead a monthly fixed fee for additional operation costs. For example, one temporary user installed a whirlpool in his backyard as he could not be charged for additional water consumption. Consequently, mediators developed legal strategies to counteract these abuses and raised the maintenance costs at the beginning (Interview 17, CEO of a for-profit mediator agency, 20.02.2015). Lastly, loaning contracts can easily and quickly be signed electronically, which reduces the administrative costs and optimizes the work efficiency of the mediator agencies in that model (Büchi & Gehrig, 2014).

*Stage 1: The Emergence of Institutionalized but Non-Commodified Temporary Housing in Zurich During the 1980s*

In Zurich, temporary housing agencies working on a non-profit basis (stage 1) are organized under the tenancy law regime (Table 5.9). Typically, the mediators rent an apartment or a whole building from the property owner through a temporary rental contract and sublet the apartments to specific target groups. These social organizations were founded more than 30 years ago and appeared as a political response to the demands of the "1980-opera-riot". They are therefore still strongly connected to the municipal government.

The investigated non-profit housing organizations "Woko" (in German: "Studentische Wohngenossenschaft"), "Juwo" (in German: "Jugendwohnnetz"), and "Domicil foundation" receive public financial support for their services and thus represent an integrated part of Zurich's social housing policy system. In other words, these social institutions are part of an approach to ensure adequate and affordable housing provisions for vulnerable and low-income households (e.g., young adults, families, and social welfare recipients) in the city. By helping them to manage their

**Table 5.9** The two stages towards the commodification of temporary housing in Zurich

| | *Stage 1—Non-commodified model* | *Stage 2—Commodified model* |
|---|---|---|
| Legal framing | Tenancy law regime—tenants are protected against unfair dismissal and rent increase | Loaning law regime—residents only pay for maintenance costs, but can be evicted at short-notice |
| Organizations studied | Woko, Juwo, Domicil foundation | Projekt Interim |
| Mediators' objectives | Professional non-profit-oriented mediators who<br>• provide their services *on the users' behalf;*<br>• take social responsibility for people with housing needs;<br>• are part of the local social housing policy strategy | Professional profit-oriented mediators who<br>• provide their services *on the owners' behalf;*<br>• work for profit;<br>• participate in housing provisions of the private profit-oriented sector |
| Property owners' objectives | Owners give their permission to use the premises. In return, they get rent as payment | Owners willingly paying for the mediators' services to maximize their planning predictability in times of urban land scarcity and pressure on the housing market |
| | *Additional benefits in both models:* protection against squatting and vandalism; moral satisfaction to contribution to affordable housing provision in times of affordable housing shortages | |
| Temporary users' objectives | Temporary housing as affordable, flexible, and centrally located opportunity. Most temporary users, however, would prefer long-term and stable housing solutions (see Sect. 9.4.4) | |

daily life (budget, housing rules, household work), they act mostly as social workers and take responsibility for the tenants' social integration and security within urban neighborhoods. The main motivation of these social organizations is to widen the housing possibilities for the specific needs of their social target groups. They select beneficiaries according to set criteria (age, income, degree of education) and provide apartments exclusively for persons in difficult living situations. As a result, the agencies identify themselves as an integrated part of the tenants' community and do not provide affordable housing to generate monetary returns (Interview 11, Director of Domicil foundation, 21.01.2015).

*Woko*, for example, was founded as a self-help association in 1956 and was later transformed into a housing cooperative in the 1970s. Its core business lies in renting affordable housing units to students. Additionally, Woko has always been a mediator between students looking for accommodation and property owners. In 1987 (after the 1980s social protests in Zurich), a public foundation for student housing SSWZ (in German: "Stiftung für Studentisches Wohnen Zürich") was created next to Woko. Thanks to SSWZ, it became possible to build new apartments for students for the first time in Zurich. While Woko focused on administration and facility management of student housing, the buildings were mainly owned by the foundation SSWZ, the municipalities of Zurich and Winterthur, the Canton of Zurich, and the Swiss Federal Institute of Technology (Interview 7, Director of Woko, 12.01.2015).

*Juwo*—the housing network for young people—is legally organized as an association and was founded as a direct response to the "1980s-social-movements" and the affordable housing crises in Zurich of that time. Juwo mainly arranges temporary housing options for a young and low-income segment of the population (Interview 4, Director of Juwo, 08.01.2015).

The private *foundation Domicil* was founded in 1994 and, as the others, follows a clear social objective. Its core interest is providing—wherever possible—long-term affordable housing. Particularly, Domicil works together with low-income people, families, single parents or people who experienced discrimination in the housing market. Many of them receive social welfare contributions. Temporary housing is only used as alternative strategy because it is very hard to find long-term affordable rental options that they can mediate to their clients in Zurich (Interview 11, Director of Domicil foundation, 21.01.2015).

> We show people how to clean a cooker, where 'Migros' [the local grocery store] is and what type of cleaning equipment they need to clean with. We also provide information about ventilation or mold prevention in the apartment or how to deal with the neighbour. We explain how to cooperate with the housekeeper and the property administration. All these different levels when it comes to housing (…). (Interview 11, director of Domicil foundation, 21.01.2015)

Since the 1980s, the affordable housing shortage has become a very strong driving force for the institutionalization of temporary housing in Zurich. It serves as a flexible and individualized approach to address specific housing needs. Hence, the non-profit agencies started to expand their regular housing portfolio with temporary apartments. In a context of land scarcity and increasing redevelopment of existing buildings, this housing type has become practiced more intensively since the beginning of the 2010s (Interview 7, Director of Woko, 12.01.2015).

> We arrange temporary housing if people are totally in emergency. If we can simply not find any other solution. [...] In Zurich, there exists a clear housing shortage for this group of people we care about. [...] It is precarious. [...] Therefore, we started to add temporary housing to our portfolio. (Interview 11, director of Domicil foundation, 21.01.2015)

Property owners who allow temporary housing mediated by the non-profit sector in their premises recognize this model either as a time and money-saving opportunity in the phase that precedes redevelopment or as an effective protection against squatting and vandalism (or both). In their view, they benefit from higher predictability, secure economic profit, and lower maintenance costs than when leaving the property empty. Interestingly, they gain higher financial income due to the rent than property owners who rely on for-profit temporary housing (see next section). Some property owners also emphasize the moral satisfaction when enabling affordable housing options for users in need and when working together with non-profit mediators (Interview 8, private property owner, 16.01.2015).

> For the whole house, which is divided into two apartments, I get a rent of 3300 francs plus 700 francs maintenance costs for water and electricity supply per month. [...] I wanted to have about the same income as if the apartments were rented long-term. [...] Because of financial reasons. I needed the income. I do not have a house in Zurich only because it is nice. (Interview 8, 69-years old private owner, 16.01.2015)

*Stage 2: The Emergence of Temporary Housing as a Business Model in the 2010s*

Commodified temporary housing is regulated under the loaning law regime. The investigated mediator agencies (two private limited companies and one joint-stock company) organize temporary housing under loaning law to earn a living out of the vacancy business. Since the year 2010, three companies positioned themselves in the profit-oriented vacancy business in Zurich (Projekt Interim GmbH, Intermezzo AG, and novac solutions GmbH). Their core business activity is managing real-estate vacancies in a profit-oriented manner on the owners' behalf and using them for temporary uses of all kind (e.g., creative economy, shared office spaces, or housing units).

> We organise temporary uses. Normally before conversion or demolition of buildings. [...] We started in 2011 and professionalised in 2013. [...] This was because we received requests. (Interview 17, CEO Projekt Interim, 20.02.2015)

The most successful firm in the vacancy business we investigated is the private firm Projekt Interim GmbH (limited company). Originally, this firm was organized as a non-profit organisation until its share-holders changed the business structure to a private profit-oriented limited company in 2013. According to the founders, they changed their business structure to for-profit as they acknowledged an increasing demand from the owners' side. In recent years, complex densification projects within the urban built environment have required longer waiting times for building permits which is why the number of property vacancies has started to escalate. This means that many property owners terminate rent contracts even though they do not already have a building permit or know the date on which they can start reconstructing. In case the rent contracts expire but no approved building permit can be demonstrated, a potential vacancy period between dismissal and demolition/renovation emerges. As a consequence, temporary use suddenly becomes all the more profitable for the initiators of Projekt Interim GmbH since owners demand professional services to manage these vacancies. The firm brings together people with a wide range of expertise such as technical skills in real-estate management, a good network in Zurich's creative scene, and legal knowledge about local and national regulations regarding temporary use (Interview 17, CEO Projekt Interim, 20.02.2015).

Property owners who choose working together with profit-oriented companies follow the clear objective to minimize risks of development delays in a city under densification pressure and a tight urban housing market. They are under high economic investment pressure and fear losing money due to building delays, which is why they need to be sure to start with the new construction on time. Their main objective is to increase predictability and flexibility in a period of economic uncertainty

(Interview 34, CEO Intermezzo AG, 26.04.2019). Therefore, they willingly pay for the mediator's services to benefit from the legal possibility of evicting the temporary users at short notice. From a financial point of view, temporary housing under loaning law mainly affects the property owners because they agree to pay for the mediators' management skills and the legal security to expel occupants (Interview 33, CEO Novac Solutions AG, 24.04.2019).

> After legal advice, we [institutional property owners] recognised that if someone with a loaning contract applies for a contract extension, the person does not have any legal possibility to extend. [...] This way, potential objections are directly off the table. At the same time, we can get temporary users out of the house within a short time. They can do nothing about it. [...] It is primarily about planning security. (Interview 14, institutional property owner, 30.01.2015)

*Effects of Letting Vs. Loaning on Temporary Users' Housing Situation*

According to a study on residents' income in the city of Zurich (Martinovits, 2014), the majority of the occupants living in the investigated temporary apartments can be identified as low-income, with a monthly net-income per person of 1000–3500 Swiss francs (Table 5.9). As a rule of thumb, in Switzerland, it is generally assumed that a quarter of the household net-income can be spent on housing costs without negatively impacting other dimensions of life (FOSI & FOH, 2015).

Under loaning law, the temporary housing prices are much cheaper than under tenancy. For instance, when renting a 4.5 room temporary apartment in Zurich (see investigated project Nr. 3: 2260 CHF for 4.5 rooms), the price is more than three times as expensive as when loaning temporary housing (see project 5: 740 CHF for 4.5 rooms) (Table 5.10). Due to these massively cheaper offers under loaning law, the socio-economic profile of temporary users changed significantly in the 2010s. Nowadays, besides the temporary users of the creative scene (e.g., students, young urban creatives) more and more working-poor families, young families, people with social aid, and low-income immigrants decide to benefit from low-cost and centrally located dwellings. In comparison to options on the regular housing market, temporary housing is offered below market rates (in letting and loaning), although in substandard conditions (e.g., with bad sanitary, insulation, and heating facilities).

> It is a very special house. The ceilings are very low. Everything is crooked. You cannot find any right angle here. If you look at this wall, you surely have 5 to 10% inclination. It has no heating. [...] It is very rudimentary. It has a shower cabin downstairs but up here it has no shower. [...] Nobody would invest money to renovate something because it is a temporary use. (Interview 2, 26-year-old male, temporary user and student at Zurich University, 06.01.2015)

In addition to affordability, some temporary users—namely students and young professionals—appreciate the flexibility provided by temporary housing. As they have not settled down yet, they acknowledge temporary living as a unique opportunity to explore the city (Interview 9, 27-years-old architect, 17.01.2015).

However, out of the seven temporary user parties investigated (three students, two young urban professionals, one working-poor family, and one low-income immigrant family) all of them stated that they would not decide to live in a temporary apartment if they could have a long-term, stable, and similarly cheap alternative on the regular housing market. It is especially difficult with children to live in temporary housing because of the frequent changes of backgrounds and friends. Temporary users living with their family also stated that they had to live in temporary apartments due to their work during night shifts. They were dependent on living at a low cost but also close to their workplace in the city center so temporary housing remained the only option (Interview 10, single mother with two children, working as cleaning assistant, 17.01.2015).

**Table 5.10** Comparable example of a 4.5 room temporary housing apartment under letting and loaning (Wüest & Partner, 2017, and questionnaire with temporary users)

| *Model* | *Project Nr.* | *Rent/loan per month for temporary apartment* | *Median market rents in the same neighborhood* |
|---|---|---|---|
| Stage 1: noncommodified model | 3 | 2260 CHF/4.5 rooms | 2530 CHF/4 to 4.5 rooms |
| Stage 2: commodified model | 5 | 740 CHF/4.5 rooms | 1850 CHF/4 to 4.5 rooms |

> At this moment, it is financially ideal. Also in Zurich, where it is almost impossible to find something for a good price at a good location. [...] But for a family, it is hard. A little tough. We are now looking for a long-term apartment because of our child. It would be nice if we could have a little rest. (Interview 13, 32-year-old male temporary user, working as cook and freelancer, living together with his wife and his one-year old boy, 28.01.2015)

Astonishingly, temporary users are financially more affected in the non profit model than they are in the commodified model. From a purely economic perspective, this is legitimized by the fact that paying for the rent ensures that their housing rights remain protected in court.

### *5.5.5 Discussion: All in Favor of Commodified Temporary Housing?*

The main research questions of this article address the tension between housing provision, temporary use, and tight urban housing markets as follows: *What are the institutional arrangements that make for-profit temporary housing possible? Which rationales of the different actors involved in this system explain its expansion?* And what are the consequences on the different categories of actors? Then, we show how institutional arrangements and the strategic behavior of different actors involved in temporary housing are at the core for answering these research questions. Not only do our results reveal how the municipal regulatory framework affects the temporary use of urban housing stocks. They also show how different actors involved in temporary use respond to and are impacted by new legal practices such as the shift towards loaning law in temporary housing.

We show that a commodified temporary housing model that is managed on the owners' behalf has emerged in Zurich. In this new business model, temporary housing is favored by a coalition of all major actors involved—property owners, mediators, temporary users, and municipal authorities. Despite the apparent agreement, however, this does not remain without consequences for those in need of affordable housing. We reveal that the short-term interests of the property owners and the for-profit mediators get the upper hand over the long-term and stable housing needs of low-income households.

*For-profit mediator agencies* take advantage of the gaps in Swiss tenancy law to offer short-term housing solutions based on loan use regulation. Individual knowledge from the non-profit temporary use sector and a strong business intuition were the drivers to institutionalize temporary housing through these new rules of the game (loaning law) and with for-profit objectives. Although for-profit mediators insist that they are aware of their social responsibility to organize temporary housing, they contribute to the weakening of tenants' security through the promotion of a housing model outside of tenancy law.

*Property owners* in the commodified model do require payment for operation costs but not rent. Due to increasingly complex inner-city redevelopment procedures and corresponding planning delays (e.g., due to objections by neighbors), the financial investment risks for owners increased in recent years. In this situation, temporary housing under loaning law fulfills a specific niche function on real-estate markets in the sense that it increases planning predictability, legal security, flexibility, and economic security for the owners. More specifically, in an environment where land is scarce and competition to use this land is increasing, property owners can no longer afford to leave their properties empty for an unpredictable duration. Inevitably, they are interested in a time and money-saving solution to remain flexible and economically competitive. In the end, because of the almost repealed protection of tenants' rights and the absence of any legal tenancy obligation to provide maintenance services, owners enjoy maximal economic security and full decision-making power.

The need for better predictability might even be reinforced by the municipal police regulation on housing squats which accepts squatting as long as landowners cannot demonstrate a formal building permit. Paradoxically, this attitude might create additional pressure on the owners as their property is no longer protected by the state in the period of vacancy and planning uncertainty. Hence, the owners are in need of finding a short-term solution which helps them to manage their vacant properties. Although the municipality of Zurich does not directly financially support the for-profit model, they contribute to its economic success by tolerating the loaning law regime in housing. Even though this model frames housing in a completely different way, it seems that the city government has not yet realized potential detrimental consequences. To counterpoise this decision, politicians and NGOs in other Swiss cities have started to call for legal prohibition of loaning law practice in housing, for instance in the city of Basel. Here, a temporary user is legally allowed to claim that the costs paid by the users for maintenance are too high for its acceptance as a loan. Instead, it can be classified as rent which makes eviction at short notice illegal (City of Basel, 2018).

Finally, the *temporary users*—namely low-income families, single parents, people receiving social aid, young urban professionals, and students—appreciate the ability to live centrally and at a low cost in Zurich. However, these vulnerable groups are caught in a vicious circle leading to dependence on precarious housing solutions and the erosion of their social rights and protection in housing. By signing loaning law contracts, temporary users abandon their legal protection as they do not have the ability to extend contracts or to claim their rights in court. Under this legal framing, they remain totally dependent on the conditions offered by the owners and mediators. In the investigated non-profit model, in contrast, temporary housing can still be interpreted as a part of the existing social housing policy system as the provision of housing is organized for the users'—not the owners'—benefit. Nevertheless, even if non-profit temporary housing follows clear social objectives, it still needs to be critically questioned whether the requested rent prices are justified for temporary apartments that are mostly substandard and designated to be demolished.

### 5.5.6 *Housing Is a Matter of Tenancy Law, Not Loaning Law!*

While there is a growing body of literature critically discussing how temporary urbanism affects urban social life (Galdini, 2019; Lai et al., 2018; Vallance et al., 2017), and another line of research focusing on the potentials of temporary use for the flexible development of urban spaces in general (Cardullo et al., 2018; Dubeaux & Sabot, 2018; Németh & Langhorst, 2014; Özdemirli, 2014), research on how different forms of temporary use affect the housing situation in cities is still thin (Lara-Hernandez et al., 2020; van Holm, 2020). More qualitative and quantitative research is therefore needed to understand how temporary housing changes everyday life in cities and potentially leads to precarious living situations for lower-income groups. Future research should focus more intensively on the social dimension of sustainability in cities (affordability, tenure security, stability, etc.) to understand how to cope with tight urban housing markets and intensifying scarcity of land.

In this article, we explained the emergence of a commodified temporary model in the Swiss urban context. Even though our results are limited to the city of Zurich, potential for generalization results from the following identified causal mechanisms which are expected to have broader significance in other urban contexts too: our analysis reveals that even though the city council publicly commits to affordable and socially equitable housing development (Municipal Constitution, Art. 2, Para. 5), in temporary housing, it assists in bypassing tenancy law by accepting the loaning law regime in housing. This, in turn, assures increased flexibility and predictability for the owners (no contract termination deadline and no corresponding judicial uncertainty). The municipal government might be reluctant to prohibit the loaning law regime for temporary housing as it might hinder property owners from densifying their parcels as investment risks increase. Simultaneously, property owners benefit from an economic incentive to raise the profit margin when obtaining densification measures due to smaller transaction costs. For temporary users, however, we see that loaning regulation leads to the erosion of their social rights, stability, and protection in housing as it promotes a precarious standard and short-term perspective of living. This model stands for a more general shift towards the acknowledgment of housing as a commodity and investment asset rather than as a basic human need and unique kind of good (Harvey, 2005, 2012; Marcuse, 1985; Rolnik, 2013). We finally identify a risk that low-income residents become increasingly excluded from inner-city housing as the supply of new housing—in particular through support measures for housing cooperatives—targets the middle class instead of the lower socio-economic segments of the population—a highly unsustainable urban development scenario! If the for-profit temporary housing model becomes more mainstream and competes even more directly with non-profit firms, it may reinforce the residents' dependence on the owners' short-term decisions and increasingly become a social challenge for the city government. In the long run, public expenses for social aid might rise as the number of residents suffering from unstable housing conditions increases and more people potentially become dependent on social welfare contributions.

Let's not open Pandora's box—housing is a matter for tenancy law, not loaning law, to protect the users' stability, security, and long-term right to housing. We are convinced that Zurich's municipal government is in the power position to change the legal setting to prohibit temporary housing under loaning law and to minimize further flexibilisation of the housing sector. As demonstrated in other Swiss cities (e.g. Basel), legal changes in tenancy law neither lead to an increase of urban vacancies nor to a prohibition of non-residential temporary uses. Instead, temporary housing vacancies are managed under non-profit objectives and with predictable tenure conditions for the users. Simultaneously, Zurich's city government should point particular attention to housing provisions for those with the lowest incomes. To look more closely at those who pay the social price of densification and corresponding urban upgrading measures is essential if urban quality and viability is to be retained for all, including more vulnerable socio-economic groups.

## References

Aalbers, M. B. (2017). The variegated financialization of housing. *International Journal of Urban and Regional Research, 41*, 542–554.

Aalbers, M. B, & Christophers, B. (2014). The housing question under capitalist political economies. *Housing, Theory and Society, 31*, 422–428.

Achtenberg, E. P., & Marcuse, P. (1986). Toward the decommodification of housing. In R. Bratt, C. Hartman, & A. Meyerson (Eds.), *Critical perspectives on housing* (pp. 474–483). Temple University Press.

Alterman, R. (1990). *Private supply of public services: Evaluation of real state exactions, linkage, and alternative land policies*. University Press.

Alterman, R. (2010). *Takings international: A comparative perspective on land use regulations and compensation rights*. ABA Publications.

Alterman, R. (2012). Land-use regulations and property values: The ‹Windfalls Capture› idea revisited. In *The Oxford handbook of urban economics and planning* (pp. 1–34). Oxford University Press.

Althaus, J., Grunwald, N., & Kreuzer, V. (2009). *Ortserkundung in der Raumplanung. Materialien zur Projektarbeit*. Institute of Urban Planning.

Amin, A., & Thrift, N. (2002). *Reimagining the Urban*. Policy Press.

Ancell, S., & Thompson-Fawcett, M. (2008). The social sustainability of medium density housing: A conceptual model and Christchurch case study. *Housing Studies, 23*(3), 423–442.

Angst, M., Klaus, P., Michaelis, T., Müller, R., & Wolff, R. (2008). *Zone*imaginaire. Argumente und Instrumente für Zwischennutzungen in der Schweiz* (pp. 1–127). irap INURA.

Arthurson, K. (2001). Achieving social justice in estate regeneration: The impact of physical image construction. *Housing Studies, 16*(6), 807–826.

Balmer, I., & Gerber, J.-D. (2017). Why are housing cooperatives successful? Insights from Swiss affordable housing policy. *Housing Studies, 33*(3), 361–385.

Barbier, E. B. (1987). The concept of sustainable economic development. *Environmental Conservation, 14*(2), 101–110.

Basel-City Council. (2009). *Areal Schoren. Ratschlag Festsetzung eines Bebauungsplans.*

Basel-City Council. (2016). *Wohnraumentwicklungsstrategie des Kantons Basel-Stadt 2012–2016.*

Basel-City Council. (2018). *Zonenplanrevision Teil II: Ratschlag an den Grossen Rat.*

Bathelt, H., & Glückler, J. (2005). Resources in economic geography: From substantive concepts towards a relational perspective. *Environment and Planning A, 37*(9), 1545–1563.

Beck, R., & Schulthess, J. (2019). Was tun gegen steigende Mieten? *Republik.* Available at: https://www.republik.ch/2019/08/07/was-tun-gegen-steigende-mieten. Accessed 7 Aug 2019.

Bibby, P., Henneberry, J., & Halleux, J. M. (2020). Under the radar? "Soft" residential densification in England, 2001–2011. *Environment and Planning b: Urban Analytics and City Science, 47*(1), 102–118.

Bishop, P., & Williams, L. (2012). *The temporary city.* Routledge.

Blomquist, W. (2012). A political analysis of property rights. In D. Cole & E. Ostrom (Eds.), *Property in land and other resources* (pp. 369–384). Lincoln Institute.

Blumner, N. (2006). *Planning for the unplanned: Tools and techniques for interim use in Germany and the United States.* Occasional Papers, pp. 1–31.

Bourassa, S., Hoesli, C., & Scognamiglio, D. (2010). Housing finance, prices, and tenure in Switzerland. *Journal of Real Estate Literature, 18*(2), 262–282.

Boyko, C. T., & Cooper, R. (2011). Clarifying and re-conceptualising density. *Progress in Planning, 76*(1), 1–61.

Bramley, G., Dempsey, N., Power, S., Brown, C., & Watkins, D. (2009). Social sustainability and urban form: Evidence from five British cities. *Environment and Planning A, 41*(9), 2125–2142.

Bramley, G., & Morgan, J. (2003). Building competitiveness and cohesion: The role of new housebuilding in central Scotland's cities. *Housing Studies, 18*, 447–471.

Bramley, G., & Power, S. (2009). Urban form and social sustainability: The role of density and housing type. *Environment and Planning B: Planning and Design, 36*(1), 30–48.

Brenner, N., Marcuse, P., & Mayer, M. (2012). *Cities for people, not for profit: Critical urban theory and the right to the city* (N. Brenner, P. Marcuse, & M. Mayer, Eds.). Routledge.

Broitman, D., & Koomen, E. (2015). Residential density change: Densification and urban expansion. *Computers, Environment and Urban Systems, 54*, 32–46.

Bromley, D. W. (1992). The commons, common property, and environmental policy. *Environmental and Resource Economics, 2*, 1–17.

Brunaupark Tenants Association. (2019a). *Interessensgemeinschaft Brunaupark.*

Brunaupark Tenants Association. (2019b). *Petition Leben im Brunaupark.*
Büchi, R., & Gehrig, F. (2014). Erstreckung ausgeschlossen: Die Gebrauchsleihe als ideale Vertragsform für die Zwischennutzung von Liegenschaften. *Jusletter.ch*, pp. 2–8.
Buitelaar, E., & Needham, B. (2007). Property rights and private initiatives: An introduction exploring alternatives to customary land use planning. *The Town Planning Review, 78*, 1–8.
Bürgin, M. (2017). Aktivierungstherapie für Immobilienpatienten. VLP-ASPAN *Raum & Umwelt, 2*, 8–14.
Burri, S. (2015). Genossenschaftswohnungen: Grosser Run, kleine Chancen. In *Swiss Radio and Television.* Available at: https://www.srf.ch/news/schweiz/genossenschaftswohnungen-grosser-run-kleine-chancen. Accessed 18 Oct 2020.
Burton, E. (2000). The compact city: Just or just compact? A preliminary analysis. *Urban Studies, 37*(11), 1969–2006.
Burton, E. (2003). Housing for an urban renaissance: Implications for social equity. *Housing Studies, 18*(4), 537–562.
Cardullo, P., Kitchin, R., & Di Feliciantonio, C. (2018). Living labs and vacancy in the neoliberal city. *Cities, 73*, 44–50.
Castells, M. (1983). *The city and the grassroots: A cross-cultural theory of urban social movements.* Edward Arnold.
Chiu, R. L. H. (2003). Social sustainability, sustainable development and housing development: The experience of Hong Kong. In R. Forrest & J. Lee (Eds.), *Housing and social change. East-West perspectives.* Routledge.
Chiu, R. L. H. (2004). Socio-cultural sustainability of housing: A conceptual exploration. *Housing, Theory and Society, 21*(2), 65–76.
City of Basel. (2018). Und sie mieten doch! In *WOZ [Die Wochenzeitung].* Available at: www.woz.ch/1842/wohnen/und-sie-mieten-doch. Accessed 03 Jan 2019.
City of Kloten. (2019). *Protokoll Stadtrat Kloten: Klotener Volksinitiative «wohnen für alle»: Antrag auf Gültigkeitserklärung und Ablehnung der Initiative.* Kloten.
City of Zurich. (2016). *Analyse Bauliche Erneuerung in Zahlen—Erneuerung von Wohnbauten in der Stadt Zürich 2000–2015.*
City of Zurich. (2019a). *Who owns Zurich?* Available at: https://www.stadt-zuerich.ch/content/prd/de/index/statistik/publikationen-angebote/publikationen/webartikel/2016-11-10_Wem-gehoert-Zurich.html. Accessed 23 July 2019.
City of Zurich. (2019b). *Bericht Energiepolitik 2017–2018.*
City of Zurich. (2020). *Bauliche Verdichtung aktuell.* Available at: https://www.stadt-zuerich.ch/content/prd/de/index/statistik/publikationen-angebote/publikationen/webartikel/2020-04-07_Bauliche-Verdichtung-aktuell.html.

Accessed 18 Oct 2020.
Colomb, C. (2012). Pushing the urban frontier: Temporary uses of space, city marketing, and the creative city discourse in 2000s. *Journal of Urban Affairs, 34*, 131–152.
COSD and CSO Zurich [Cantonal Office for Spatial Development and Cantonal Statistical Office]. (2014). *Akzeptanz der Dichte.*
Credit Suisse Zurich. (2020). *Brunaupark Projekt*. Available at: https://brunaupark.ch/faq/. Accessed 25 Mar 2020.
Daneshpour, A., & Shakibamanesh, A. (2011). Compact city; does it create an obligatory context for urban sustainability? *International Journal of Architectural Engineering & Urban Planning, 21*(2), 9–23.
Davidson, M. (2010). Social sustainability and the city. *Geography Compass, 4*(7), 872–880.
Davy, B. (2005). Grundstückswerte, Stadtumbau und Bodenpolitik. *Vhw Forum Wohneigentum, 2*, 67–72.
Davy, B. (2012). *Land policy: Planning and the spatial consequences of property*. Routledge.
Dawkins, C. J., & Nelson, A. C. (2002). Urban containment policies and housing prices: An international comparison with implications for future research. *Land Use Policy, 19*, 1–12.
De Decker, P. (2009). Social rental agencies: Still a splendid idea? *European Journal of Homelessness, 3*, 215–230.
Dembski, S., & Salet, W. (2010). The transformative potential of institutions: How symbolic markers can institute new social meaning in changing cities. *Environment and Planning A, 42*, 611–625.
Dempsey, N. (2010). Revisiting the compact city? *Built Environment, 36*, 5–8.
Dempsey, N., Bramley, G., Power, S., & Brown, C. (2011). The social dimension of sustainable development: Defining urban social sustainability. *Sustainable Development, 19*, 289–300.
Demsetz, H. (1967). Towards a theory of property rights. *The American Economic Review, 57*(2), 347–359.
Dubeaux, S., & Sabot, E. (2018). Maximizing the potential of vacant spaces within shrinking cities, a German approach. *Cities, 75*, 6–11.
Elsinga, M. G., Hoekstra, J. S. C. M., Sedighi, S. M. A., & Taebi, B. (2020). Toward sustainable and inclusive housing: Underpinning housing policy as design for values. *Sustainability, 12*(5), 1–15.
Fainstein, S. F. (2010). *The just city*. Cornell University Press.
FDF [Federal Deparment of Finance], FOE [Federal Office for Energy], and FOH [Federal Office for Housing]. (2009). *Steuerliche Anreize für energetische Sanierungen von Gebäuden Studie der interdepartementalen Arbeitsgruppe.*

Feldges, V. (2019). *Städtebauliche Verträge.* Lecture CUREM CAS Urban Management Zurich.

Florida, R. (2002). *The rise of the Creative Class.* Basic Books.

Fluri, K. (2017). *Postulat 16.4028 Schweizer Ortsbilder erhalten.*

FOC [Federal Office for Culture]. (2018). *Schweizer Ortsbilder erhalten Bericht des Bundesrates in Erfüllung des Postulates.*

FOE [Federal Office for Energy]. (2015). *Wirkung steuerlicher Anreize für energetische Gebäudesanierungen und mögliche Hemmnisse bei deren Finanzierung Zusammenfassung.*

FOE and FOH [Federal Office for Energy and for Housing]. (2015). *Energetische Sanierung. Auswirkungen auf Mietzinsen.*

FOE [Federal Office for Energy]. (2018). *Das Gebäudeprogramm.*

FOF, FOT, FOE and FOH [Federal Offices for Finance, Taxes, Energy and Housing]. (2009). *Steuerliche Anreize für energetische Sanierungen von Gebäuden Studie der interdepartementalen Arbeitsgruppe.*

FOH [Federal Office for Housing]. (2012). *Preisgünstiger Wohnraum mittels raumplanerischer Massnahmen? Wirkungsanalyse und Handlungsempfehlungen.*

FOH [Federal Office for Housing]. (2014a). *Bericht über die Ergebnisse der Vernehmlassung zur Änderung des Mietrechts im Obligationenrecht (OR).*

FOH [Federal Office for Housing]. (2014b). *Vorkaufsrecht der Gemeinden – Bericht zuhanden des Bundesrates Zusammenfassung Ausgangslage.*

FOH [Federal Office for Housing]. (2014c). *Wohnungspolitischer Dialog Bund, Kantone und Städte Zwischenbericht der Arbeitsgruppe.*

FOH [Federal Office for Housing]. (2016a). *Bericht Energetische Gebäudesanierungen im Mietwohnungsbereich.*

FOH [Federal Office for Housing]. (2016b). *Wohnungspolitischer Dialog Bund, Kantone und Städte Zwischenbericht der Arbeitsgruppe.*

FOH [Federal Office for Housing]. (2017). *Gemeinnütziges Wohnen im Fokus. Ein Vergleich zu Miete und Eigentum*, Grenchen.

FOH [Federal Office for Housing]. (2018). *Verzeichnis Formularpflicht für den Anfangsmietzins gemäss Artikel 270 Absatz 2 OR.*

FOH [Federal Office for Housing]. (2019). *Zusammenhang zwischen Einkommens- und Energiearmut sowie die Folgen energetischer Sanierungen für vulnerable Gruppen. Eine qualitative Analyse.*

FOSD [Federal Office for Spatial Development]. (2016). *ISOS und Verdichtung. Bericht der Arbeitsgruppe.* Bern.

FOSD [Federal Office for Spatial Development]. (2017). *Bauzonenstatistik Schweiz 2017. Statistik und Analysen.* Available at: https://www.are.admin.ch/dam/are/de/dokumente/grundlagen/dokumente/bauzonenstatistik-schweiz-2017-statistik-und-analysen.pdf.download.pdf/Bauzonenstatistik_Schweiz_2017.pdf. Accessed 18 Oct 2020.

FOSD [Federal Office for Spatial Development]. (2018). Webpage Federal Office for Spatial Development. Available at: https://www.are.admin.ch/are/de/home.html. Accessed 22 Aug 2019.

FOSI [Federal Office for Social Insurances] and FOH [Federal Office for Housing]. (2015). *Wohnungsversorgung in der Schweiz—Bestandesaufnahme über Haushalte von Menschen in Armut und prekären Lebenslagen.*

FSO [Federal Statistical Office]. (2015). *Bevölkerunsstand- und Struktur.* Neuchatel.

FTA [Federal Tax Administration]. (2017). *Umsetzung der steuerlichen Massnahmen im Gebäudebereich geht in die Vernehmlassung.*

Galdini, R. (2019). Urban re-use practices in contemporary cities: Experiences in Europe. *Cities, 87*, 103–105.

Gennaio, M. P., Hersperger, A. M., & Bürgi, M. (2009). Containing urban sprawl-evaluating effectiveness of urban growth boundaries set by the Swiss Land Use Plan. *Land Use Policy, 26*, 224–232.

George, A. L., & Bennett, A. (2005). *Case Studies and Theory Development in the Social Sciences.* MIT Press.

Gerber, J.-D. (2018). A Swiss perspective on time-limited property rights: Strategic use for active land policy. In J.-D. Gerber, T. Hartmann, & A. Hengstermann (Eds.), *Instruments of Land Policy—Dealing with Scarcity of Land* (pp. 260–264). Routledge.

Gerber, J.-D., Hartmann, T., & Hengstermann, A. (2018). *Instruments of land policy-dealing with scarcity of land* (J.-D. Gerber, T. Hartmann, & A. Hengstermann, Eds.). Routledge.

Gerber, J.-D., Koepfel, P., Nahrath, S., & Varone, F. (2009). Institutional Resource Regimes: Towards sustainability through the combination of property-rights theory and policy analysis. *Ecological Economics, 68*(3), 798–809.

Gerber, J.-D., Nahrath, S., & Hartmann, T. (2017). The strategic use of time-limited property rights in land-use planning: Evidence from Switzerland. *Environment and Planning A, 49*(7), 1684–1703.

GFOBRP [German Federal Institute for Research on Building, Urban Affairs and Spatial Development]. (2016). *Mietrecht und energetische Sanierung im europäischen Vergleich.* Bonn.

Hall, P., & Taylor, R. (1996). Political science and the three new institutionalisms. *Political Studies, 44*, 936–957.

Hardegger, T. (2017). *Förderung des gemeinnützigen oder preisgünstigen Wohnungsbaus. Vorkaufsrecht für Gemeinden.*

Hardin, G. (1968). The Tragedy of the Commons. *Science, 162*(3859), 1243–1248.

Hardin, G. (1991). The tragedy of the unmanaged commons: Population and the disguises of providence. *Trends in Ecology & Evolution, 9*(5), 175–199.

Harloe, M. (1982). Towards the decommodification of housing?: A comment on council house sales. *Critical Social Policy, 2*(4), 39–42.
Harloe, M. (1995). *The people's home: Social rented housing in Europe and America*. Blackwell.
Hartmann, T., & Spit, T. (2015). Dilemmas of involvement in land management—Comparing an active (Dutch) and a passive (German) approach. *Land Use Policy, 42*, 729–737.
Hartmann, T., & Gerber, J.-D. (2018). Land, scarcity, and property rights. In J.-D. Gerber, T. Hartmann, & A. Hengstermann (Eds.), *Instruments of Land Policy* (pp. 3–7). Routledge.
Harvey, D. (2005). *A brief history of neoliberalism*. Oxford University Press.
Harvey, D. (2012). *Rebel cities: From the right to the city to the urban revolution*. Verso.
Healey, P., & Barrett, S. M. (1985). *Land policy: Towards a research agenda*. Gower.
Healey, P. (1996). The communicative turn in planning theory and its implications for spatial strategy formation. *Environment and Planning B: Planning and Design, 23*(2), 217–234.
Healey, P. (1999). Institutionalist analysis, communicative planning, and shaping places. *Journal of Planning Education and Research, 19*(2), 111–121.
Healey, P. (2007a). The new institutionalism and the transformative goals of planning. In N. Verma (Ed.), *Institutions and planning*. Elsevier.
Healey, P. (2007b). *Urban complexity and spatial strategies: Towards a relational planning for our times*. Routledge.
Heeg, S. (2013). Wohnungen als Finanzanlage. Auswirkungen von Responsibilisierung und Finanzialisierung im Bereich des Wohnens. *zeitschrift sub/urban*, Heft 1, pp. 75–99.
Hengstermann, A., & Gerber, J.-D. (2015). Towards active land policy—The new Swiss Spatial Planning Act. *fub, 6*, 241–250.
Hengstermann, A. (2019). *Von passiver Bodennutzungsplanung zu aktiver Bodenpolitik—Eine Fallstudie zur Wirksamkeit von bodenpolitischen Instrumenten anhand von Lebensmittel-Discountern* (1st ed.). Bern.
Holm, A., & Kuhn, A. (2011). Squatting and urban renewal: The interaction of squatter movements and strategies of urban restructuring in Berlin. *International Journal of Urban and Regional Research, 35*(3), 644–658.
Holman, N., Mace, A., Paccoud, A., & Sundaresan, J. (2015). Coordinating density; working through conviction, suspicion and pragmatism. *Progress in Planning, 101*, 1–38.
Hood, C. (1983). *The tools of government*. Macmillan Education.
Hood, C., & Margetts, H. (2007). *The tools of government in the digital age*. Palgrave Macmillan.

Huber, M. (2020). Bewilligung für umstrittene Brunaupark-Neuüberbauung. *Tages-Anzeiger*. Accessed 18 Mar 2020.

IFHP [International Federation for Housing and Planning]. (2019). *The Social Cities Index: A tool for urban leaders*. Denmark.

Jenks, M., Burton, E., & Williams, K. (1996). *The compact city: A sustainable urban form?* (1st ed.). Routledge.

Jessop, B. (2001). Institutional re(turns) and the strategic-relational approach. *Environment and Planning A, 33*, 1213–1235.

Jonkman, A. (2020). Patterns of distributive justice: Social housing and the search for market dynamism in Amsterdam. *Housing Studies*. https://doi.org/10.1080/02673037.2020.1739232

Kadi, J., & Ronald, R. (2014). Market-based housing reforms and the "right to the city": The variegated experiences of New York, Amsterdam and Tokyo. *International Journal of Housing Policy, 14*(3), 268–292.

Kébir, L. (2010). Pour une approche institutionnelle et territoriale des ressources. Bruxelles: Peter Lang. In M. Maillefert, O. Petit, & S. Rousseau (Eds.), *Ressources, patrimoine, territoires et développement durable* (pp. 69–86). P.I.E. Peter Lang.

Khan, M. A. (1995). Sustainable development: The key concepts, issues and implications. *Sustainable Development,* (3), 63–69.

Knoepfel, P. (1986). Distributional issues in regulatory policy implementation—The case of air quality control policies. In A. Schnaiberg, N. Watts, & K. Zimmermann (Eds.), *Distributional conflicts in environmental policy* (pp. 363–379). Gower.

Knoepfel, P., Larrue, C., Varone, F., & Hill, M. (2011). *Public policy analysis*. Policy Press.

Knoepfel, P., Csikos, P., Gerber, J.-D., & Narath, S. (2012). Transformation der Rolle des Staates und der Grundeigentümer in städtischen Raumentwicklungsprozessen im Lichte der nachhaltigen Entwicklung. *Nomos Verlagsgesellschaft, 53*(3), 414–443.

Knoepfel, P., Nahrath, S., & Varone, F. (2007). Institutional regimes for natural resources: An innovative theoretical framework for sustainability. In P. Knoepfel (Ed.), *Environmental policy analyses. Learning from the past for the future. 25 Years of research* (pp. 455–506). Springer.

Korthals Altes, W. (2016). Forced relocation and tenancy law in Europe. *Cities, 52*, 79–85.

Kriesi, H., Koopmans, J., Duyvendak, W., & Guigni, M. (1995). *New social movements in Western Europe*. UCL Press.

Krueger, R., & Agyeman, J. (2005). Sustainability schizophrenia or "actually existing sustainabilities?" Toward a broader understanding of the politics and promise of local sustainability in the US. *Geoforum, 36*(4), 410–417.

Lai, L. W. C., Chau, K. W., & Cheung, P. A. C. W. (2018). Urban renewal and redevelopment: Social justice and property rights with reference to Hong Kong's constitutional capitalism. *Cities, 74*(6), 240–248.

Lara-Hernandez, J. A., Coulter, C. M., & Melis, A. (2020). Temporary appropriation and urban informality: Exploring the subtle distinction. *Cities, 99*, 1–10.

Laur, F. (2019). Das Geisterhaus am Schorenweg. *Basler Zeitung*, August 27, 21pp.

Lawson, J. (2009). The transformation of social housing provision in Switzerland Mediated by Federalism, Direct Democracy and the Urban/Rural Divide. *European Journal of Housing Policy, 9*(1), 45–67.

Lees, L. (2008). Gentrification and social mixing: Towards an inclusive urban renaissance? *Urban Studies, 45*(12), 2449–2470.

Lees, L., Slater, T., & Wyly, E. (2008). *Gentrification*. Routledge.

Lefebvre, H. (1991). *The production of space*. Blackwell.

Linder, W. (1994). *Swiss Democracy: Possible Solutions to Conflict in Multicultural Societies*. St. Martin's Press.

Littig, B., & Griessler, E. (2005). Social sustainability: A catchword between political pragmatism and social theory. *International Journal Sustainable Development, 8*, 65–79.

liveinguardians.com. (2020). liveinguardians online: http://liveinguardians.com. Accessed 27 Jan 2020.

Locke, T. (1689/1977). *Labor theory of ownership*. In Second Treatise of Government. Walter Euchner.

Maissen, F. (2018). Rendite um jeden Preis? Massenkündigungen von Mietwohnungen nehmen zu. *Swiss Radio and Television*. Available at: https://www.srf.ch/news/schweiz/rendite-um-jeden-preis-massenkuendigungen-von-mietwohnungen-nehmen-zu. Accessed 22 July 2019.

Mandelbaum, S. J. (1985). The institutional focus of planning theory. *Journal of Planning Education and Research, 5*, 3–9.

Manzi, T. (2010). *Social sustainability in urban areas: Communities, connectivity and the urban fabric*. Earthscan.

Marcuse, P. (1985). To control gentrification: Anti-displacement zoning and planning for stable residential districts. *Yearbook of Construction Articles* (pp. 931–952). New York University Review of Law and Social Change.

Marcuse, P. (2012). A critical approach to solving the housing problem. In N. Brenner, P. Marcuse, & M. Mayer (Eds.), *Cities for people, not for profit*. Routledge.

Marcuse, P. (2016). Gentrification, Social Justice and Personal Ethics. *International Journal of Urban and Regional Research, 39*(6), 1263–1269.

Martel, A. (2020). Ein Wohnparadies für alle? Weshalb das nicht funktionieren kann. *Neue Zürcher Zeitung*. Accesses 27 June 2020.

Martínez, M. A. (2013). The Squatters' movement in Europe: A durable struggle for social autonomy in urban politics. *Antipode, 45*(4), 866–887.
Martinovits, A. (2014). *Mietzinse und Einkommen. Eine Auswertung der Schweizerischen Arbeitskräfteerhebung (SAKE) für die Stadt Zürich.*
Marx, K. (1859). *Zur Kritik der politischen Ökonomie.* Karl Marx/Friedrich Engels - Werke. (Karl) Dietz Verlag.
Marx, K. (1868). *Über die Nationalisierung des Grund und Bodens.* Karl Marx/ Friedrich Engels - Werke. (Karl) Dietz Verlag.
Mayer, M. (2007). Contesting neoliberalism of urban governance. In H. Leitner, J. Peck, & E. S. Sheppard (Eds.), *Contesting neoliberalism: Urban frontiers* (pp. 90–115). Guilford Press New.
McKenzie, S. (2004). *Social sustainability: Towards some definitions.* University of South Australia.
Mitlin, D. and Satterthwaite, D. (1996). Sustainable development and cities. In C. Pugh (Ed.), *Sustainability, the Environment and Urbanisation* (pp. 23–62). Earthscan Publications Limited.
Mulliner, E., Smallbone, K., & Maliene, V. (2013). An assessment of sustainable housing affordability using a multiple criteria decision making method. *Omega, 41*, 270–279.
Murphy, K. (2012). The social pillar of sustainable development: A literature review and framework for policy analysis. *Sustainability: Science, Practice, & Policy, 8*(1), 15–29.
Nahrath, S. (2018). A Swiss perspective on pre-emption rights: Impact without application. In J. D. Gerber, T. Hartmann, & A. Hengstermann (Eds.), *Instruments of land policy—Dealing with scarcity of land* (pp. 213–217). Routledge.
Nebel, R., Hollenstein, K., Di Carlo, G., Niedermaier, M., & Scholl, B. (2017). *Schweizweite Abschätzung der Nutzungsreserven 2017.* ETH University.
Needham, B., Buitelaar, E., & Hartmann, T. (2018). *Planning law and economics—The rules we make for using land* (2nd ed.). Routledge.
Needham, B., & Verhage, R. (1998). The effects of land policy: Quantity as well as quality is important. *Urban Studies, 35*(1), 25–44.
Németh, J., & Langhorst, J. (2014). Rethinking urban transformation: Temporary uses for vacant land. *Cities, 40*, 143–150.
Nicol, L. A., & Knoepfel, P. (2008). Institutional regimes for sustainable collective housing stocks. *Swiss Political Science Review, 14*(1), 157–180.
North, D. C. (1994). Economic performance through time. *The American Economic Review, 84*(3), 359–368.
Oppliger, M. (2016). In der Schorenmatte zeigt Basel, wie planvolle Verdichtung aussehen könnte. *Tageswoche.*
Ostrom, E. (1990). *Governing the commons: The evolution of institutions for collective action.* Cambridge University Press.

Ostrom, E. (2007). Institutional rational choice: An assessment of the institutional analysis and development framework. In P. Sabatier (Ed.), *Theories of the policy process* (pp. 21–64). Westview Press.

Ostrom, E. (2009). A general framework for analyzing sustainability of social-ecological systems. *Science, 325*(5939), 419–422.

Oswalt, P., Overmeyer, K., & Misselwitz, P. (2003). *Urban Pioneers: Temporary Use and Urban Development in Berlin.* Jovis.

Özdemirli, Y. K. (2014). Alternative strategies for urban redevelopment: A case study in a squatter housing neighborhood of Ankara. *Cities, 38*, 37–46.

Pérez, F. (2020). 'The miracle of density': The socio-material epistemics of urban densification. *International Journal of Urban and Regional Research, 44*(4), 617–635.

Polese, M., & Stren, R. (2000). *The social sustainability of cities. Diversity and the management of change* (M. Polese & R. Stren, Eds.). University of Toronto Press.

Priemus, H. (2011). Squatters and municipal policies to reduce vacancy, Evidence from The Netherlands. In *Enhr Conference 2011* (pp. 1–6).

Pruijt, H. (2003). Is the institutionalization of urban movements inevitable? A comparison of the opportunities for sustained squatting in New York City and Amsterdam. *International Journal of Urban and Regional Research, 27*(3), 133–157.

Pruijt, H. (2013). The logic of urban squatting. *International Journal of Urban and Regional Research, 37*, 19–45.

Rawls, J. (1972). *A theory of justice.* Harvard University Press.

Rérat, P., & Lees, L. (2011). Spatial capital, gentrification and mobility: Evidence from Swiss core cities. *Transactions of the Institute of British Geographers, 36*(1), 126–142.

Rérat, P. (2012). The new demographic growth of cities: The case of reurbanisation in Switzerland. *Urban Studies, 49*, 1107–1125.

Rolnik, R. (2013). Late neoliberalism: The financialization of homeownership and housing rights. *International Journal of Urban and Regional Research, 37*(3), 1058–1066.

Rousseau, J.-J. (1762). *Du contrat social ou principes du droit politique.* H. Denhardt.

Salamon, L. M. (2000). The tools of government: A guide to the new governance. *Fordham Urban Law Journal, 28*(5), 1611–1675.

Savini, F., Majoor, S., & Salet, W. (2015). Dilemmas of planning: Intervention, regulation, and investment. *Planning Theory, 14*(3), 296–315.

Scally, C. P., & Tighe, J. R. (2015). Democracy in Action?: NIMBY as Impediment to Equitable Affordable Housing Siting. *Housing Studies, 30*(5), 749–769.

Schmid, S. (2020). Die Schweiz und ihre Mietpreis-Hotspots. *Republik*, pp. 1–4.

Schoop, F., Kälin, A., & Hofer, K. (2020). Das lange Warten auf den Abbruch. *Neue Zürcher Zeitung*, pp. 18–19.
Slaev, A. (2016). Types of planning and property rights. *Planning Theory, 15*, 23–41.
Smith, N. (2017). *The new urban frontier*. Routledge.
Stahel, T. (2006). *Wo-Wo-Wonige!—Stadt- und wohnpolitische Bewegungen in Zürich nach 1968*. University Press.
Statistics Basel-City. (2019). *Wanderungsbefragung 2018. Ergebnisbericht.*
Statistical Office City of Basel. (2019). *Bevoelkerungsszenarien.*
Statistical Office City of Basel. (2020). *Bevölkerungsentwicklung Stadt Basel, Bevölkerungsentwicklung*. Available at: https://www.statistik.bs.ch. Accessed 18 Oct 2020.
Statistical Office City of Zurich. (2017). *Immer mehr Leerkündigungen bei Umbauten*. Available at: https://www.stadt-zuerich.ch/prd/de/index/statistik/publikationen-angebote/publikationen/webartikel/2017-09-14_Immer-mehr-Kuendigungen-bei-Umbauten.html. Accessed 26 Mar 2020.
Statistical Office City of Zurich. (2020). *Bevölkerungs- und Wirtschaftsentwicklung Stadt Zürich*. Available at: https://www.stadt-zuerich.ch/prd/de/index/statistik/themen/bevoelkerung.html. Accessed 25 June 2020.
Stone, M. E. (2006). What is housing affordability? The case of the residual income approach. *Housing Policy Debate, 17*(1), 151–184.
Sturzenegger, M. (2020). UNO-Sonderberichterstatterin eilt Zürcher Mietern zu Hilfe. *Tages-Anzeiger.* Accessed 9 Jan 2019.
Swiss Federal Council. (2014). *Botschaft zu einem Rahmenkredit für Eventualverpflichtungen in der Wohnraumförderung für die Jahre 2015–2021 vom 20. August 2014* (pp. 6441–6460).
Swiss Federal Council. (2016a). *15.044 Geschäft des Bundesrates zur Änderung des OR. Mietrecht.*
Swiss Federal Council. (2016b). *Strategie Nachhaltige Entwicklung 2016–2019.*
Swiss Federal Council. (2017). *Verdichtetes Bauen in Ortszentren fördern, aber wie? Bericht des Bundesrates in Erfüllung des Postulats von Graffenried 14.3806 vom 24. September 2014.*
Swiss Federal Council. (2018). *Schweizer Ortsbilder erhalten. Bericht des Bundesrates in Erfüllung des Postulates.*
Tenants Association Basel. (2019). *Einsprache gegen Bauprojekt Schorenweg.*
Theurillat, T., & Crevoisier, O. (2013). The sustainability of financialized urban megaproject: The case of Sihlcity in Zurich. *International Journal of Urban and Regional Research, 37*(6), 2052–2073.
Theurillat, T., Rerat, P., & Crevoisier, O. (2014). The real estate markets: Players, institutions and territories'. *Urban Studies, 52*(8), 1414–1433.

Touati-Morel, A. (2015). Hard and soft densification policies in the Paris city-region. *International Journal of Urban and Regional Research, 39*(3), 603–612.

Townroe, P. M. (1996). Urban sustainability and social cohesion. In C. Pugh (Ed.), *Sustainability, the environment and urbanization* (pp. 101–150). Earthscan.

UN [United Nations] Habitat. (2016). *Habitat III—United Nations Conference on Housing and Sustainable Urban Development* (pp. 1–23). Quito.

Vallance, S., Dupuis, A., & Edwards, S. (2017). Temporary use and the onto-politics of 'public' space. *Cities, 70*, 83–90.

Vallance, S., Perkins, H. C., & Dixon, J. E. (2011). What is social sustainability? A Clarification of Concepts. *Geoforum, 42*(3), 342–348.

van der Krabben, E., & Jacobs, H. (2013). Public land development as a strategic tool for redevelopment: Reflections on the Dutch experience. *Land Use Policy, 30*(1), 774–783.

Van Holm, E. J. (2020). Evaluating the impact of short-term rental regulations on Airbnb in New Orleans. *Cities, 104*, 1–9.

Viallon, F.-X. (2018). Added value capturing in Switzerland: How much is enough? In J.-D. Gerber, T. Hartmann, & A. Hengstermann (Eds.), *Instruments of Land Policy—Dealing with Scarcity of Land* (pp. 59–68). Routledge.

von Graffenried, A. (2014). *14.3806 Verdichtetes Bauen in Ortszentren fördern, aber wie?*

Vonmont, A. (2016). *Sanieren mit dem Blick fürs Ganze.*

Wehrmüller, A. (2014). *TENLAW: Tenancy law and housing policy in multi-level Europe, National Report for Switzerland* (pp. 1–176). European Union.

Weingaertner, C., & Moberg, Å. (2014). Exploring social sustainability: Learning from perspectives on urban development and companies and products. *Sustainable Development, 22*(2), 122–133.

Williams, K., Burton, E., & Jenks, M. (2000). *Achieving sustainable urban form* (K. Williams, E. Burton, & M. Jenks, Eds.). E & FN Spon Press.

Williamson, O. E. (2000). The new institutional economics: Taking stock, looking ahead. *Journal of Economic Literature, 38*(3), 595–613.

Woodcraft, S. (2012). Social sustainability and new communities: Moving from concept to practice in the UK. *Procedia—Social and Behavioral Sciences, 68*, 29–42.

Wüest & Partner. (2017). Zwischennutzungen: Ein Instrument gegen Leerstände? *Immo-Monitoring*, pp. 73–83.

Yin, R. K. (2018). *Case study research and applications. Design and methods* (6th ed.). Sage.

Yung, B. & Lee, F., (2012). Right to housing in Hong Kong: Perspectives from the Hong Kong community. *Housing, Theory and Society, 29*(4), 401–419.

Zukin, S. (2009). *Naked city: The death and life of authentic urban places*. Oxford University Press.
Zurich City Council. (2013). *Gerechter. Die Entwicklung der Bau- und Zonenordnung der Stadt Zürich.*
Zurich City Council. (2017). *Programm Wohnen. Stadtrat von Zürich.*
Zurich City Council. (2019a). *Auszug aus dem Protokoll des Stadtrats. GR Nr. 2019/200.*
Zurich City Council. (2019b). *Der Stadtrat von Zürich an den Gemeinderat. GR Nr. 2019/90.*
Zurich Municipal Parliament. (2019). Motition der Fraktionen SP, Grüne und AL des Gemeinderats Zürich GR. Nr. 2019/90. Stadtverwaltung Zurich.

CHAPTER 6

# Discussion of Results

## 6.1 Discussion of Key Results

In this chapter and Chapter 7, a final discussion and conclusion are drawn *whether* and *how* the implementation of densification objectives leads to social sustainability in housing. This is done by analyzing the institutional regime (key finding 1), and the different actors' strategies using housing stocks under scarce urban land use conditions (key findings 2 to 4). Making explicit the local governance mechanisms of possible trade-offs *and* power games among actors (key finding 5) in densification procedures is a new contribution of this research project to neoinstitutionalist political ecology research.

> **SQ1**: How do institutional rules affect the outcomes of densification in terms of social sustainability in housing?

**Key empirical finding 1: Institutional incoherence between landowners' and tenants' rights, and the extent of rules influencing the implementation of densification objectives, lead to socially unsustainable housing development**

Results of Articles 1 to 4 *confirm hypothesis 1* (see Sect. 3.6): in a context of densification, the Swiss institutional regime in force is indeed key for disentangling *how* and *why* social sustainability in housing is produced. Article 1 demonstrates *how* the unequal distribution of legal



G. Debrunner, *The Business of Densification*,
https://doi.org/10.1007/978-3-031-49014-9_6

power between public policy *and* property rights in the Swiss federal regulatory framework, *but also* the extent of formal rules impacting the densification implementation process, affect the way actors use housing goods and services (e.g. affordable living space, energy source). Whereas property titles are strongly secured by the Swiss Constitution *and* the Civil Code, and are very enduring and long-lasting, Swiss public policies regulating housing stocks are quickly revised depending on current political majorities. More precisely:

Article 1 shows that while in Swiss federal planning, tax and energy policy the parliament has only recently agreed to introduce new legislation that stronger enshrines energy efficiency objectives, legal amendments to support social interests of densification (e.g. in housing, planning, or social welfare policy) have not been made. Federal policy instruments in favor of socially sensitive redevelopment such as the introduction of a fixed minimum share of 10% non-profit housing in the Swiss Constitution[1] or the obligation for private homeowners to pass tax benefit for energetic renovation to tenants, have been politically debated but never fully enacted.

In addition, on November 20, 2020, the Swiss Federal Court only recently decided that landlords can increase the rent by several hundred Swiss francs in case new renters move into an apartment (in stock as well as in new built housing). This decision was made based on the grounds that institutional owners (e.g. pension funds) would make too little profit due to the low interest rates on the property markets. It is therefore expected that the number of mass dismissals in the rental segment will increase in Switzerland since owners can benefit of fast and increasing rent revenue (Perricone, 2020).

Consequently, even though the land use context in Swiss cities has changed, Swiss property owners still benefit from public subsidies for energetic renovation or modernization wherever housing stock is being densified. Owners must neither pass the received subsidies to the tenants nor are they obliged to stick to certain rent levels where real estate is demolished and refurbished resulting in higher costs for residents. In

[1] On February 9, 2020, the Swiss electorate voted on the popular initiative "more affordable homes". The initiative proposed that across Switzerland at least 10% of newly built homes should belong to non-profit developers (mostly housing cooperatives). Swiss voters rejected the initiative. Final results show it won 42.9% of the vote nationwide, with 57.1% against, on a turnout of 41.7% (Swiss Federal Council, 2020).

sum, owners are encouraged to densify through public incentives (e.g. for energetic renovation, energy efficiency), but without any restrictions or duties regarding rent increase, dismissal of elderly, or disruption of long-enduring communities. While this power guaranteed to landowners has not triggered any conflicts on unbuilt green- or brownfields, the legal tensions between owners' and residents' interests have increased in a context of densification.

Indeed, even though the socio-spatial conditions under the compact city model have changed significantly, results show that property owners are still allowed to densify existing housing stocks without further tenancy restriction. They can terminate an open-ended rent contract within three months without any specific reasons and without any legal obligation to compensate the tenants, for example, for moving costs. Moreover, in newly built housing, it is possible for landowners to start the rent at a new level. They do not have to follow rent levels as would be mandatory for renovated real estate. Therefore, property owners are highly motivated to densify through complete demolition as they can make a lot of profit through increased rent revenue after rebuilding.

Moreover, their position is secured by strongly protected property titles. This means that the owner is free to define the profit margin to be targeted on the parcel and can set the rents according to market prices. Hence, urban densification is essentially profit-driven in Switzerland since property owners do not face any institutional boundaries that would prevent them from acting socially unsustainable. Housing providers are not forced by law to follow a more socially inclusive solution (e.g. in terms of housing affordability, prevention of discrimination).

Tenants, in turn, are not in the legal position to be heard or to counteract. Regardless of their strength of social integration, age, or years of residency in a neighborhood, tenants can legally be evicted within three months. Under the Swiss rules of tenancy, residents do also not need to be informed about upcoming redevelopment tasks *before* receiving the contract termination. In daily practice, this often leads to social eviction at short notice. Tenants may counteract in court; however, in most cases they do not use this option as they neither have the financial means nor the expert knowledge to do so.

Under this given legal setting, results of Article 1 show that the Swiss federal government withdraws from its responsibility of covering the housing needs for all income segments. In the name of federalism, it passes the duty on to the cantons, municipalities, and ultimately the individual subjects. The federal government does not enforce more rigorous institutional rules and policy instruments to protect tenants. Because legal regulations and policy instruments, which would directly alleviate the land rent (e.g. rent level or social eviction controls, quotas for affordable housing, zoning measures etc.), would lead to strong owners' resistance (see key finding 2). Simultaneously, owners make full use of their rights to private ownership including their rights to control housing access, rights of management, rights of exclusion, or rights of alienation.

Ultimately, results of Articles 2 and 4 reveal that this institutional incoherence between property owners' and tenants' rights in the Swiss federal regulatory framework trickles down to cantonal and municipal levels. While private property owners' profit-making interests are very strongly protected by law in Switzerland, tenants do not experience the same legal protection of their affordable housing needs. This legal dependence from the owners' position encourages the performance of densification as an "Eco-Business" at the expense of its social function (key finding 5).

**SQ2**: What use strategies do actors (owners and non-owners) follow to contribute to socially sustainable housing in a densifying city?

**Key empirical finding 2: Resistance power of landowners prevents socially sustainable housing development in a densifying city**

Results of the four articles of this book *confirm hypothesis 2* (Sect. 3.6): private property owners together with the private building and real estate industries working on their behalf are identified to be the main driving force behind tenants' social exclusion in Swiss cities. For them, densification represents a lucrative business and financial asset since they benefit from enhanced revenue on centrally located parcels (more apartments). In comparison to the development on greenfield outside of the built-up urban areas, they regard residential urban densification as an attractive business for stable rates of return and as a solid risk diversification strategy. As a consequence, consolidation in Swiss cities increasingly becomes a process led by capital in the sense that it is primarily shaped and guided by private actors. They intend to boost the city's economic activity through intensive investment into housing (re)development. Owners (including speculators), investors, and the private construction industry congruently benefit from this emerging "Business of Densification" as it leads to a substantial increase in land rent and income revenue.

Moreover, the private companies (investors, banks, insurances) studied in Articles 2 and 4 are of a certain size, and are thus able to possess large-scale housing projects. They can manage them through considerable means (purchase of land, demolition of old infrastructure, construction costs, etc.). In addition, investors (such as Credit Suisse pension fund) acknowledge the compact city as an opportunity to legitimize the transfer of property rights power over public planning principles (densification) from one scale to another and to justify the precedence of densification accorded to financial values over use values. They promote densification in a way that neglects distributional consequences or social equity issues and ignores the residents' needs for increased affordability, integration into decision-making, or community cohesion. Results show that these firms neither engage with the concerned communities nor are they aware of intersectional power structures. Because their rights to private property prevent them from choosing more socially stable solutions.

Goals of social inclusion, community cohesion, or housing accessibility are not considered in decision-making since investors *by law* do not feel compelled to act socially responsibly (key finding 1). As long as they are not legally required to do so, and their position is secured by strong property titles, owners acknowledge no need to support tenants' interests. Furthermore, results of Article 1 show that the interests of these private development firms are well represented in the federal parliament. Profit-oriented goals of densification become increasingly integrated into political games taking place at higher decision-making levels. Through the coalition of political forces (in legislative and executive committees), the introduction of formal rules that aim to promote more socially inclusive objectives in housing is constantly prevented, for example, in regard to the provision of affordable housing, mass dismissals, or improved tenure security.

**Key empirical finding 3: Effective tenants' resistance against rent increase and displacement is limited due to their weak legal standing in front of powerful landowners**

The assumption made in *hypothesis 3 cannot be confirmed*: results of the four articles of this book reveal that tenants confronted with contract termination in most cases neither have the financial means nor the legal standing to counteract the landlords' decision effectively. Under the given Swiss federal institutional setting (key finding 1), *and* the decision-making strategies applied by property owners (key finding 2), tenants' capacity for resistance against rent increase and displacement is limited. Swiss property owners are well equipped to resist tenants' claims thanks to their strong position as titleholders. Even though residents try to resist through street rallies, formal petitions, or the collaboration with local tenants' associations, they do not succeed in defending their social interests effectively (e.g. for affordable housing provision). They face discrimination due to their low-income status and weak legal position. Hence, even in the Swiss direct democratic system, tenants' grassroots resistance does not lead to enhanced socio-political pressure on governments and owner-actors to obtain measures against dismissal and displacement.

In the studied municipalities, NIMBY-efforts prevent planned projects only to the extent that new developments might be delayed, but never fully rejected. Moreover, property owners (but also public authorities) develop strategies to counteract such NIMBY-opposition *in advance* in order to prevent building delays, which cost them a lot of money. For instance, Article 4 shows that property owners have started to promote a temporary housing model that works outside of tenancy law (loaning law), which obliges low-income households to live in precarious housing situations in buildings shortly before demolition. The city government of Zurich has not prohibited this practice even though it leads to the constant erosion of tenants' social rights and legal security. In doing so, owners prevent former residents from squatting their buildings, which would delay their consolidation and upgrading plans. By supporting this housing format, public authorities manage to implement densification goals effectively.

In addition, results of Articles 2 and 4 reveal that property owners have developed professional communication strategies with tenants and planning authorities to hinder social resistance. For example, to convince residents to participate, investors develop strategies to make densification palatable to them even though it leads to the loss of their social habitats. They offer, for instance, economic compensation measures to tenants such as alternative dwellings in parts outside the city to diminish NIMBYism that may delay their projects. Simultaneously, public planning administrations too develop a financial language and management skills that smooth the dialogue with private investors *and* tenants. Participatory forms of planning such as cooperative planning, test planning, or urban design competitions play a central role in this context as they make the densification process more efficient and profitable for both sides—investors *and* local authorities.

In essence, the constant erosion of their social rights, relations, and contacts to family members, neighbors, and friends worsens the housing situation of tenants living in private rental housing stocks in Swiss cities. Even though they do their utmost to rebel and to make their protest against densification and upgrading visible, their weak legal standing hampers effective resistance. Even though they might be included in participative forms of planning, in case they do not agree with the plans, they do not have the legal standing to actually incorporate their visions into formal rules and regulation. As a consequence, they remain heavily dependent on the decisions made by the landowners.

Those being displaced have become the victims of powerful forces of capitalist urbanization and differential spending power in the Swiss legal system. To cope with this legal situation, informal mechanisms of accountability occur within self-organized arrangements that enable community-based collective action, for example, in the form of non-profit housing associations or non-profit temporary housing organizations (see Article 4). However, I argue in this book that current housing challenges need to be solved in a way that includes profit-oriented housing providers as well, because the non-profit sector only holds a small share of the total housing stock in Switzerland.

Finally, results of Article 1 show that enhanced tenants' lobbying power is needed to change the regulatory framework in force in favor of more social goals. While the Swiss homeowners' association manages to constantly influence the implementation of energy policy instruments or tenancy regulations to make densification less socially diverse, for many years, tenants' activist groups have not succeeded in activating new rules that promote tenants protection. This lack of access to formal decision-making at the federal level (due to lack of lobbying power) constantly reinforces social inequalities in the Swiss legal and housing policy system. This subsequently also affects regulatory regimes at the cantonal and municipal levels.

**Key empirical finding 4: Municipal authorities' strategic activation of public and private law instruments leads to socially sustainable housing development in a dense city**

Results of Article 3 *confirm the assumption made in hypothesis 4*: a shift towards active municipal land policy is necessary to promote socially sustainable housing in a context of urban densification. Municipal planning administrations are in the key position to intervene into private development interests. They can prevent tenants' social exclusion, but they therefore must know *how* to densify.

Article 3 provides a central theoretical contribution to this field of research in the sense that—by conducting comparative case study research—the article makes a clear suggestion what an effective municipal land policy strategy for socially sustainable housing transformation might look like. In a context of urban land scarcity, public authorities do not only need to understand the technical challenges such as contamination. They must also (re)consider neighborhood conflicts or the underlying property rights' structure if they aim to fight urban sprawl without discrimination, displacement, and social exclusion. Article 3, moreover, indicates that planners should be aware that, at the center of the action system are the owners, who initiate the densification process as well as the private homebuilding industry working on their behalf. Together they form a private alliance that favors densification in order to promote business (key finding 2).

To counteract such profitability objectives—resulting in rising rents and displacement—results in Article 3 show that municipal planning authorities must find ways to deal with the power of titleholders. More specifically, they need to understand *how* to activate specific public and private law instruments that do not always need to limit property owners' rights, but also work *with* property rights. Such an active approach requires public intervention outside of a purely market-based investment logic. Besides the mere introduction of new policy instruments (e.g. quota for affordable housing), it involves the strategic activation of existing formal rules.

Moreover, Article 2 demonstrates that to legitimize such land acquisition strategy, broad political support, knowledge, as well as financial and personnel resources are needed. These are not always given, even in big and wealthy cities such as Zurich or Basel. Consequently, making densification more socially sensitive is a matter of the ability of municipal planning administrations to familiarize themselves with the entire range of existing intervention possibilities. The strengthening of more social measures must be given more weight in the planning process, and should be considered and included more actively into formal decision-making processes (for details see Sect. 7.1).

**SQ3**: How does the implementation of densification objectives impact social sustainable housing outcomes?

**Key empirical finding 5: The Business of Densification—the basic human need for shelter is being neglected in favor of economic-environmental ("Eco-Business") functions of housing estates**

The final assumption made in *hypothesis 5 can be confirmed*: results of all four articles show that even though densification has been introduced as a legally binding policy objective in order to promote sustainable settlement development effectively, its implementation process is far from socially sustainable in the sampled Swiss cases. In the Swiss context, densification objectives are implemented in a way that neglects the social pillar of sustainability in favor of economic (profit) and environmental (energy efficiency) dimensions. Consolidation is promoted as an "Eco-Business" by coupling urban competitiveness with ecologic viability goals, while neglecting social aspects such as housing affordability, -availability, or tenants' inclusion into formal decision-making. More precisely:

Results of Articles 1 to 4 reveal that densification—as a key objective of Swiss planning policy—only gets implemented if private titleholders agree to obtain new development. To convince owners to participate in densification policy efforts, public authorities agree to the development terms and conditions dictated by the landowners. Because public planning administrations cannot trigger densification objectives alone. They are reluctant to intervene against property owners using measures that directly influence the land rent (e.g. stricter zoning) because they fear strong resistance. Therefore, planning authorities implement densification according to the landowners' profitability terms rather than not being able to promote consolidation goals at all. Otherwise, they would risk that densification as key element of environmentally sustainable settlement transformation might slow down—or even come to a standstill—if property owners refuse its implementation.

Under the flag of "sustainable urban development", this results in consolidation implemented in the form of "green gentrification" (Budd et al., 2008: 266), which couples ecologic modernization (energy efficiency) with densification goals, but at the expense of a city's social qualities. The policy shift towards densification enables owners to (re)develop existing real estate through the construction of more and newly renovated apartments at the same location (key finding 2). They can increase rent revenue steadily and substantially. Moreover, even though public authorities would be the ones responsible for counteracting trends of gentrification through targeted government intervention, results of Articles 2 to 4 show that the political instrumentalization of the compact city is a very real possibility for municipalities.

Indeed, public authorities use the compact city as a cover to push forward policies and interventions that have goals other than social stability. For example, energy efficiency or climate adaption objectives. Since the compact city model fits into very different political agendas, this increases the possibility that it is used instrumentally. Densification enables governments to grow, attract, and accommodate new (potentially higher-income) residents. Hence, authorities try to justify the (re) development of old housing stocks with the need to produce increased net financial capacity for managing urban growth. In the city of Zurich, for example, municipal planning authorities define planning documents that facilitate the (re)construction of housing stocks as they are interested in attracting wealthy taxpayers in order to increase tax revenue (Articles 2 to 4). Public authorities (on behalf of local executive and legislative committees) entail an economistic perspective—yet one often integrated with ecological discourse—by highlighting the financial cost of sprawl and by mobilizing arguments in favor of dense and compact cities. Such arguments include, for instance, a reduction of transaction costs through spatial proximity, which subsequently may lead to increased urban livability.

Consequently, more than ever before, public actors incorporate gentrification into densification policy objectives—used either as a justification to obey market forces and private sector entrepreneurialism, or as a tool to direct market processes in the hopes of reorganizing urban landscapes in a more benevolent fashion. Swiss public authorities (particularly planning administrations) seem to regard gentrification as one of the social costs of economic success or ecologic renovation that is worth tolerating. To do so—in the sampled Swiss cases—they argue that failure to allow new private buildings will create definite and severe impediment to economic growth or climate protection. In Articles 2 and 3, for instance, municipal authorities legitimize their actions with the argument that the impact of doing nothing for economic or environmental sustainability would be devastating. Primarily, because "doing nothing" could potentially create additional pressure for production, jobs, and residents to move out of the city, which could foster urban decline. Terms like "partnerships", "participation", "collaboration", or "sustainability" are used instrumentally to reinforce the power of the central state or to presage the hegemony of property owners' lobbies.

While the displacement of lower income encouraged in the course of implementing densification programs is generally made a taboo, the furthering of participation and articulation options for desired groups and newcomers is widely discussed and politically promoted. However, such an urban development strategy is far from the original aims of the compact city and considered highly unsustainable as it supports social displacement and eviction of residents. In the Swiss context, results of the four articles show that implementing densification goals not only leaves the very housing situation of the lowest-income groups at risk but also that of the middle class. Through consolidation and upgrading, affordable housing units are demolished and rebuilt with higher densities but also with the consequence of higher rents. Findings reveal that social ties in neighborhoods are being disrupted at short notice because evicted tenants cannot afford rental prices on the free market after contract termination anymore. Social aspects such as housing affordability, housing availability, or tenants' access to decision-making are bypassed during the process of upgrading. Social qualities that are a city's basis of community-based initiatives and solidarity-creating capacities are jeopardized, which leads to social exclusion, gentrification, and -polarization of lower income (particularly old-aged), but also middle class. In Swiss cities, this results in a way of implementing densification goals that is highly unattainable, especially regarding its socioeconomic consequences for the residents.

## 6.2 Intermediate Summary

In summary, results so far show that the conditions for socially sustainable housing development in Swiss cities have changed in recent years due to a situation of urban land scarcity. The shift towards densification has strongly influenced actors' use interests in housing and the way they activate the regulatory institutional regime in force. At the local level, coordination among competing housing uses may be improved **through mutual adjustment of institutions** (stronger protection of tenants' rights, see key finding 1) or **strategic capacity building of municipal planners** (through active land policy, see key finding 4). However, results show that coordination in favor of socially sustainable densification is often impeded by the following:

1. *Key empirical finding 1*: the enacted federal regulations and policy instruments hinder the reduction of rent levels throughout all institutional levels (also cantonal and municipal). Findings show that the Swiss institutional regime is characterized by strong institutional fragmentation—characterized by contradictions not only *between* public policies (e.g. energy, planning, social welfare), but also between public policies *and* property rights (that are strongly protected in Switzerland). This failure in introducing more socially inclusive policy measures (e.g. quota for affordable housing, stricter tenure protection in case of modernization) is triggered by a general policy shift towards the promotion of renewable energy production regardless of its social costs ("Eco-Business");
2. *Key empirical finding 2*: the owner-actors' ability to strategically resist against public densification or against tenants' NIMBY-efforts through strategic rule activation (of property rights) or rule formulation (less tenure security). Moreover, owner-actors unite their forces and policy resources (e.g. through networks, capital, knowledge, or law) in order to prevent a backdrop of an overarching rule that could potentially lead to more social requirements in densification projects; and
3. *Key empirical finding 3*: the residents and tenants' activist groups weak legal standing in front of powerful landowners. They often fail to anchor their concerns in political documents and legislation in a targeted manner. Legislative changes in favor of stronger tenants' protection can only be amended in unique "windows of opportunity" (e.g. in times of severe housing crises) in which tenants experience broad public and political support.
4. *Key empirical finding 4*: a lack of knowledge, financial capacity, and personnel of public planning administrations (at the federal and municipal level) to resist against owners' interests and to strategically activate existing or introduce new formal rules to prevent social exclusion. Enforcing a change in the institutional setting may help tenants and public actors to reduce the rival housing uses and to boost more accurately social sustainability objectives in housing.

Chapter 7 closes with the findings answering the main research question.

## References

Budd, W., Lovrich, N., Pierce, J. C., & Chamberlain, B. (2008). Cultural sources of variations in US urban sustainability attributes. *Cities, 25*(5), 257–267.

Perricone, M. (2020). *Die Mieter haben gerade ziemlich viel Geld verloren.* https://www.srf.ch/news/schweiz/bundesgerichts-entscheid-die-mieter-haben-gerade-ziemlich-vielgeld-verloren. Accessed 15 Dec 2020.

Swiss Federal Council. (2020). Popular Initiative 'more affordable homes. https://www.admin.ch/gov/en/start/documentation/votes/20200209/iniziativa-popolare--piu-abitazioni-a-prezzi-accessibili-.html. Accessed 15 Dec 2020.

CHAPTER 7

# Final Conclusion: Governance Mechanisms for Socially Sustainable Urban Densification

Since many decades, planning and policy efforts have increased dramatically that embrace densification as a key element of sustainable settlement transformation. The compact city model has been introduced as a global attempt to incorporate green growth objectives (e.g. energy efficiency) to the level of cities (e.g. Elkin et al., 1991; Frey, 1999; Newman & Kenworthy, 1999). However, this book has demonstrated that when the concept is applied to practice, the compact city solution starts to lose some of its gloss.

Land use claims in dense city areas appear diverse and contradictory. Decision-making procedures, for example, those meant to reduce affordable housing shortages, are complex due to intricate and changing small-scale ownership structures, veto rights controlled by power actors, and intertwined interests. To capture these use conflicts and power games among actors, this book has applied a neoinstitutionalist political ecology analysis approach—the one of the Institutional Resource Regimes (IRR)—that combines theories of policy analysis (planning as a public policy), new institutional economics (property rights), *and* political ecology (power). It appraises densification as a political field by analyzing the socio-political determinants of socially sustainable housing provision and discusses arising challenges in a more solution—and practical planning-oriented manner.

More precisely, by mobilizing the IRR conceptual framework (Gerber et al., 2009, 2020; Knoepfel & Nahrath, 2007), causal relationships



G. Debrunner, *The Business of Densification*,
https://doi.org/10.1007/978-3-031-49014-9_7

between *housing as a resource*, *institutions* (both on the public policy and property-titles level), and *involved actors' strategies* are explained. In contrast to other neoinstitutional analysis attempts (e.g. Healey, 2007; Needham et al., 2018; Ostrom, 1992), such an approach enables the researcher to recognize that many different resource users can come into conflict with each other and puts particular emphasis on the distinction between the characteristics of public policy and property rights. It allows for a systematic analysis to examine *how* various actors behave in response to a specific socio-political setting. The IRR moreover manages to address issues of power, scale, politics, embeddedness, and social justice in interactions between humans and their environment.

The leading question of this book, which is—**What governance mechanisms lead to socially sustainable housing development in a dense city?**—can be answered as follows: In cities characterized by scarce land use conditions, social sustainability in housing can be achieved if local governance mechanisms are to be improved by the following:

1. *by counterbalancing* the weakness of federal and cantonal policies (particularly of planning, energy, and tenancy laws) that neglect the social pillar of sustainability in housing. This is to be done;
2. *by introducing* and/or *activating* more socially effective municipal policy instruments such as:

   - public control mechanisms *of housing finance capital* (e.g. municipal housing foundations, public subsidies for non-profit housing associations);
   - public control mechanisms *of private land* (e.g. restrictive zoning in favor of social criteria, provision of building leases to housing cooperatives, or public land acquisition);
   - *social protection mechanisms* for tenants (e.g. eviction controls, rent controls, legal protection from redevelopment or modernization).

   Such policy measures (for details see Article 3) would foster stronger protection of rents and residents from market-dependencies and thus lead to more decommodified forms of housing.
3. *by counteracting* the decision-making capacity and resistance power of private property owners and other owner-actors (e.g. private real estate industry), who are in the legal position to resist. To do so, municipal authorities must approach an active land policy strategy

that promotes decommodification of housing stocks. Besides the activation of existing or the introduction of new policy instruments (see previous argument), such a strategy includes city authorities' capability and sensitivity to promote affirmative action as well as equitable resource allocation in order to raise political pressure and to limit profiteering. Public planning administrations must take the socioeconomic consequences of densification seriously and start to plan for those with less financial means.

Closer analysis of the shift towards active municipal land policy is presented in Sect. 7.1.

This book concludes that the emerging "Business of Densification" in Swiss urban areas has city—specific implications for the integration of the urban poor and middle-classes. Insecurity of land tenure compounded by high prices and scarcity of land results in precarious housing forms such as profit-oriented temporary housing based on loaning law. Social criteria (e.g. social mixing, tenure security, housing affordability) are put in the background, while economic and ecological criteria become more prioritized. However, an urban structure in which only high-income people can continue to afford to live ultimately reduces fair distribution of, and access to life-sustaining resources (such as housing as a basic human need). If densification is approached only through a process of green gentrification (energetic modernization leading to social eviction), city sustainability will be put at risk. It cannot be achieved by supporting particular economic and environmental aspects at the cost of the social. The diminishing of one dimension affects the others.

In other words, sustainable settlement transformation calls for "the continuous creation and re-creation of adequate patterns of social organization, within which technological progress can unfold properly, the use of natural resources can be managed soundly, and the social actors of development can participate, both individually and collectively, and can share the goals and benefits of development" (Cernea, 1993: 19). To contribute to such change, this book has addressed a gap in IRR research (e.g. Balmer & Gerber, 2017; Nicol, 2013; Nicol & Knoepfel, 2008) by more closely connecting housing challenges to land scarcity, and land policy debates (e.g. Davy, 2012; Kolocek, 2017), as well as social sustainability concerns (e.g. Bramley et al., 2009; Burton, 2000; Chiu, 2004).

The foregoing argument has shown that, unless strategic municipal governmental action is taken, residents and non-profit tenants' associations will continue to remain excluded from the emerging "Business of Densification". Taking into account future challenges of land scarcity that currently evolve in many cities, the findings of this research may help municipal planners, practitioners, and policymakers to counteract trends of rising commodification in housing, and to develop new forms and modes of housing resource management in order to (re)organize paths of capital accumulation. Only by doing so, city governments will be able to adequately address social equity issues and the needs of the disadvantaged in a context of intergenerational resource stability both in the short and in the long term.

## 7.1 Policy Recommendations: Towards Active Municipal Land Policy for Socially Inclusive Urban Housing in Densifying Cities

This book explores how in daily planning practice every society sets the boundaries where commodification begins and where it ends. It reveals that where the boundaries lie is a matter of contention. In housing, the role of institutions becomes crucial for explaining exclusion and unjust societal structures. Moreover, it demonstrates that the redefinition and redistribution of the rights of private property and the profit rate derivative entails a revolution in political-economic practices. Political struggles towards decommodification, and even of freedom itself, move center-stage in the search for alternatives.

The four articles that constitute this book confirm that strategies of decommodification exist even in Switzerland, a state representing the very core of advanced capitalist economies. Even though results confirm that marketization and commodification of housing are not going to go away—what urban practitioners, local politicians, civic organizations, and other public and private stakeholders must find are more collective forms of governance and housing production so that densification processes respond to the needs of the public at large rather than simply catering to private individuals and firms.

To accomplish this aim, municipal planning authorities, *first*, can promote **the introduction of new policy instruments** to make densification more socially sensitive. Such instruments include, for instance, a quota for affordable housing, public subsidies for non-profit housing cooperatives, or social eviction controls (Article 3). However, the introduction of new rules presents by no means a panacea. Instead, making densification more socially sustainable is strongly influenced by the strategic ability of public administrators to familiarize themselves with all existing intervention possibilities.

*Second*, effort should therefore be made **to properly activate existing instruments in force**, for instance, zoning regulations. In the case of Switzerland, some municipalities (e.g. Zurich, Köniz) have indeed allocated plots of their land to non-profit foundations or cooperatives on favorable terms. By following the cost rent principle, these social organizations offer dwellings at lower prices than those determined by the free market. In Basel, moreover, municipal authorities have started to negotiate the terms and conditions of (re)development via public–private-partnerships (contracts). Such action could become more generalized to integrate social equity goals more effectively into concrete densification projects. Planning authorities need to initiate a more project-oriented approach (rather than plans only) and collaborative approaches (rather than hierarchical organizations) to cope with social challenges. However, it must also be noted that there are risks associated with project planning as well. The most important one is the issue of equality of treatment of different stakeholders involved in the planning process—which local planning administrations need to be aware of when entering a new project.

*Third*, making densification more socially sensitive presents a matter of political will for social sustainability. To promote institutional change or to legitimize a proactive land acquisition strategy in favor of tenants' social inclusion, broad political support is needed. However, local politicians often regard offending private investors' plans as too risky for the municipality's financial situation. In such situations, it takes **all the finesse and professional competencies** of municipal planning administrations and politicians (e.g. expert knowledge, financial resources, networks, personnel) to promote social aspects, because landowners have the power to defend the status quo with strong veto rights. So, even though community cohesion and residential stability are widely acknowledged as important components for urban livability, there is still a risk of downplaying this aspect in daily densification practice. Here, municipal authorities are in a key position to take responsibility in order to include local ideas about community stability and cohesion and to encode them into land use regulations.

*Fourth*, municipal planning needs **to be sensitive** to the potentially disruptive impacts of densification on local identity and diverse ways of living. This book advocates for planners and policymakers to consider social sustainability criteria (e.g. affordability, cohesion) in order to address more accurately potential trade-offs between economic, environmental, *and* social concerns of densification. Planning processes should not be limited to actors with the right of appeal but should include all affected actors such as residents of different age, income, or nationality. To ensure the inclusion of local knowledge and inhabitants' social mix, municipal planners should encourage owners to share and to discuss ideas about upcoming projects or dismissal trials *before* owners submit the building application so that formal facts and procedures are not created beforehand.

*Finally*, it must be noted that, even though local planners are theoretically able to intervene into strong market forces, the decision is not only up to them. Their action depends on whether there is political will for such an intervention strategy, which is—particularly in the Swiss liberal context—often not the case. This said, Sect. 7.1 has outlined for municipal authorities four possible intervention ways *how* to intervene strategically and actively into housing densification procedures. It therefore demonstrates *how* it might be worth at least a try to follow these paths for creating a more socially sensitive and inclusive city.

## 7.2 Theoretical and Methodological Limitations

This book has some theoretical and methodological limitations that need to be addressed in order to critically reflect upon the results. From a *theoretical perspective*, a challenge in using the IRR was to uncover the strategies behind specific actions taken and to make them visible. Depending on the institutional background, stakeholders are considered to have different means to either change these dimensions or to influence others to do so. An interdependent relationship between the actors and the institutional structure is assumed although this interaction is sometimes limited (particular in regard to informal networks). To counteract these ambiguities, my work mostly followed a policy instrument-focused approach to make the applied strategies graspable in the format of the tools activated. The way actors aimed to achieve particular goals became visible by analyzing *how* and *why* they activated specific formal rules (e.g. zoning or property rights).

Moreover, the social sustainability indicators introduced in this book (e.g. in Article 2) are considered a useful approach to evaluate housing use conflicts in dense cities. By developing a deeper understanding for the multi-faced processes that housing under scarce land use conditions is confronted with, my attempt was to connect consolidation with ongoing debates of social exclusion and gentrification. In addition, my goal was to connect the concept of social sustainability with the IRR framework that allows for a systematic analysis of social challenges in dense urban environments.

However, it must be noted that the indicators introduced do not fully present the complete picture. Each process is only a part of the complex whole constituting the city. Great attention is needed for the transferability of such norms and perspectives of social sustainability and its physical adequacy. The evaluative model provides potential for the application to, for example, other households living in existing housing stocks. Nevertheless, the extent to which the criteria can be compared to other cases with different policy and housing market contexts needs to be assessed. Within the wider Swiss situation, for instance, authentic commitment to social sustainability in housing would need to be reflected in national and local approaches and directives that move beyond rhetoric. While this research project attempts to provide a theoretical basis for a more comprehensive land use planning policy, more work will need to be performed to empirically apply the proposed evaluative criteria so as to evaluate different policy instruments from various stakeholders' perspectives.

Conclusory, by applying the IRR analytical framework, this book has evaluated social challenges in dense urban environments by combining concepts from public policy analysis (planning as a public policy), new institutional economics (property rights), and political ecology (power). Making explicit the local governance mechanisms of possible sustainability trade-offs *and* power games among actors in densification procedures is a new contribution of this research project to neoinstitutionalist political ecology research. Indeed, the differentiated discussion of the results shows that combining these concepts has led to new insights in critical environmental studies as: **(a)** power structures were systematically detected by adding a neoinstitutionalist perspective to political ecology research, *and* (**b**) power structures were explicitly discussed rather than as an integral part of public policy analysis.

This combination has led to results showing that depending on the institutional background, stakeholders have different means to either change the formal rules in force or try to influence others to do so. These governance mechanisms are at the same time the channels through which stakeholders exercise power. However, further research is needed to refine these concepts (e.g. by integrating a Foucauldian or feminist political ecology perspective) to enable analyzing *how* exactly stakeholders make use of these "power channels", and *what* factors increase their effectiveness in relation to their institutional background means.

From a *methodological perspective*, the case studies employed in Articles 1–4 provide in-depth insights into actors' behavior and regulatory regimes shaping residential densification. Even though results are limited to Switzerland (and findings should only be generalized to other cases with caution), potential for generalization to other states and cities results from the identified causal mechanisms (key findings 1 to 5), the relevance of which is expected to be broader than in the analyzed cases only. Through the clear description of the research setting, the reasons for case selection, the deductively developed research approach, as well as the critical discussion of the methods conducted (Chapter 4), the validity of the results of this book is clearly outlined.

## 7.3 Implications for Future Research: Densification, The IRR, and Beyond

This book raises further interesting questions. Even though densification has become a core objective of policy agendas across the globe, critical analysis of its socio-political limitations, challenges, and contradictions, particularly concerning its impacts on tenants from a social sustainability perspective, was largely missing when this research project started. It would therefore be interesting to initiate an **international comparative study** in order to compare the Swiss case with other examples in different state contexts. For example, to investigate *if* and *how* housing stocks are affected by densification measures in other urban contexts in order to understand *how* the process is negotiated and governed in different state settings.

Further research is indeed needed to analyze **the burgeoning field of research on land policy in planning**. In particular, *how* different stakeholders make use of certain institutional rules or policy instruments, and what factors increase their effectiveness in defending their interests. While much literature so far has focused on individual policy instruments (e.g. urban growth boundaries), the strategic combination of different instruments to reach particular densification goals still needs to be empirically further analyzed and theoretically conceptualized.

For example, it would be interesting to focus more on the role of the investors and *how* they recognize existing power relations and institutional mechanisms that regulate their access to housing in densifying cities. Decisions on the acquisition and sale of parcels, housing and building stocks, or building rights are primarily made by the head management of large investment firms. Mainly in the person of the portfolio manager, but under the supervision of the board members. To promote meaningful engagement with diverse local stakeholders and interests, it is necessary for owners to become more socially responsible. However, the question still remains *how* exactly and in *what kind* of setting. The operations, motivations, and funding structures of housing investment companies should therefore be analyzed in greater detail. Actions should also be designed to encourage each type of landownership (public, cooperative, private) to use their room for maneuvering to promote social sustainability in housing.

By applying the IRR analytical framework, this book has tried to evaluate these questions by using concepts from public policy analysis (planning as a public policy), new institutional economics (property rights), and political ecology (power), as well as by applying a qualitative case study methodology. Further cross-fertilization between different academic stands and methodological approaches can offer new insights into ongoing debates such as multi-level governance or rural densification. For instance, a more quantitative approach to evaluating socio-spatial differences, room stress of households, or socioeconomic effects of densification could provide inspiration for further study of such issues as what are the housing needs for different groups or individuals.

Moreover, it is imperative that qualitative empirical research be brought to bear on issues, such as the extent to which there are disproportionate housing risks borne by specific groups of people. Actions taken on behalf of environmental sustainability—such as the policy shifts towards the compact city model—have a consequence on the environment *and* the people. Especially but not exclusively the disadvantaged. More must be learned about those effects so that residents who are likely to be overlooked become more included into decision-making procedures. Because one thing we have learned in this book: only a socially inclusive city is also a sustainable city.

## References


Balmer, I., & Gerber, J.-D. (2017). Why are housing cooperatives successful? Insights from Swiss affordable housing policy. *Housing Studies, 33*, 361–385.

Bramley, G., Dempsey, N., Power, S., Brown, C., & Watkins, D. (2009). Social sustainability and urban form: Evidence from five British cities. *Environment and Planning A, 41*(9), 2125–2142.

Burton, E. (2000). The compact city: Just or just compact? A preliminary analysis. *Urban Studies, 37*(11), 1969–2006.

Cernea, M. M. (1993). Culture and organization—The social sustainability of induced development. *Sustainable Development, 1*(2), 18–29.

Chiu, R. L. H. (2004). Socio-cultural sustainability of housing: A conceptual exploration. *Housing, Theory and Society, 21*(2), 65–76.

Davy, B. (2012). *Land policy: Planning and the spatial consequences of property*. Routledge.

Elkin, T., McLaren, D., & Hillman, M. (1991). *Reviving the city: Towards sustainable urban development*. Friends of the Earth.

Frey, H. (1999). *Designing the city: Towards a more sustainable form*. E&FN Spon.

Gerber, J.-D., Koepfel, P., Nahrath, S., & Varone, F. (2009). Institutional resource regimes: Towards sustainability through the combination of property-rights theory and policy analysis. *Ecological Economics, 68*(3), 798–809.

Gerber, J.-D., Lieberherr, E., & Knoepfel, P. (2020). Governing contemporary commons: The institutional resource regime in dialogue with other policy frameworks. *Environmental Science and Policy, 112*(6), 155–163.

Healey, P. (2007). The new institutionalism and the transformative goals of planning. In N. Verma (Ed.), *Institutions and planning*. London.

Knoepfel, P., & Nahrath, S. (2007). Environmental and spatial development policy. In U. Klöti, P. Knoepfel, & H. Kriesi (Eds.), *Handbook of Swiss politics* (pp. 705–733). NZZ Verlag.

Kolocek, M. (2017). *The human right to housing in the face of land policy and social citizenship: A global discourse analysis*. Palgrave Macmillan.

Needham, B., Buitelaar, E., & Hartmann, T. (2018). *Planning law and economics—The rules we make for using land* (2nd ed.). Routledge.

Newman, P., & Kenworthy, J. (1999). *Sustainabily and cities: Overcoming automobile dependence*. Island Press.

Nicol, L. A., & Knoepfel, P. (2008). Institutional regimes for sustainable collective housing stocks. *Swiss Political Science Review, 14*(1), 157–180.

Nicol, L. A. (2013). *Sustainable collective housing: Policy and practice for multi-family dwellings*. Routledge.

Ostrom, E. (1992). Governing the commons: The evolution of institutions for collective action. *Natural Resources Journal, 32*(2), 415–430.

# Appendix

Annex: List of interviewees

| *Nr.* | *Date* | *Role* | *Profile of the interviewee* |
|---|---|---|---|
| 1 | 06.01.2015 | Temporary user | Female, 25 years, student |
| 2 | 06.01.2015 | Temporary user | Male, 26 years, student |
| 3 | 07.01.2015 | Cooperative property owner | "Baugenossenschaft Eidgenössisches Personal", cooperative property owner |
| 4 | 08.01.2015 | Director of a non-profit mediator agency | In German: "Jugendwohnnetz"—the housing network for young people |
| 5 | 10.01.2015 | Temporary user | Male, 32 years, employee in tourism industry |
| 6 | 12.01.2015 | Temporary user | Male, 27 years, architect |
| 7 | 12.01.2015 | Director of a non-profit mediator agency | In German: "Studentische Wohngenossenschaft"—non-profit housing cooperative |
| 8 | 16.01.2015 | Private property owner | Male, 69 years, individual property owner |
| 9 | 17.01.2015 | Temporary user | Male, 25 years, student |
| 10 | 17.01.2015 | Temporary user | Single mother with two children, 42 years, employee in cleaning industry |
| 11 | 21.01.2015 | Director of a non-profit mediator agency | Domicil foundation |
| 12 | 28.01.2015 | Cooperative property owner | "Baugenossenschaft Limmattal", coopera- tive property owner |

(continued)



G. Debrunner, *The Business of Densification*,
https://doi.org/10.1007/978-3-031-49014-9

(continued)

| *Nr.* | *Date* | *Role* | *Profile of the interviewee* |
|---|---|---|---|
| 13 | 28.01.2015 | Temporary user | Male, 32 years, working as cook and freelancer, living together with his wife and his one-year-old boy |
| 14 | 30.01.2015 | Institutional property owner | "Sansara", institutional property owner |
| 15 | 30.01.2015 | Real-estate management firm | Employee and team leader in the private firm "Livit " (joint stock company) |
| 16 | 20.02.2015 | Local homeowners' association (NGO) | Employee and expert in housing issues |
| 17 | 20.02.2015 | CEO of a for-profit mediator agency | Projekt Interim (limited company) |
| 18 | 20.02.2015 | Institutional property owner, Helvetia insurance | Institutional property owner |
| 19 | 12.03.2015 | Private property owner | Female, 39 years, private property owner |
| 20 | 06.05.2015 | Politician of the local legislative parliament | Member of the green party, expert in temporary use and housing |
| 21 | 27.08.2018 | Federal Office for Spatial Planning, Department for Sustainable Settlement Development | Head of Department |
| 22 | 29.08.2018 | Federal Office for Housing | Director |
| 23 | 13.08.2018 | Federal Office for Energy, Department for Sustainable Housing | Employee at the Department for Sustainable Energy in Housing |
| 24 | 28.08.2018 | Swiss homeowners' association | Director |
| 25 | 14.08.2018 | Swiss tenants' association | Director |
| 26 | 27.08.2018 | Swiss association of the building industry | Director |
| 27 | 05.09.2018 | Swiss association of institutional investors | Director |
| 28 | 14.08.2018 | Member of national council | Member of the Green Party |

(continued)

(continued)

| *Nr.* | *Date* | *Role* | *Profile of the interviewee* |
|---|---|---|---|
| 29 | 15.08.2019 | City of Kloten, Head of Planning Department | City planner |
| 30 | 17.09.2018 | Member of national council, President of the Swiss Cities' Association | Member of the Liberal Party (FDP) |
| 31 | 18.04.2019 | Zurich local tenants' association (NGO) | Head of the association and politician in the local legislative parliament |
| 32 | 18.04.2019 | Politician of the local legislative parliament | Member of the alternative party, expert in housing issues |
| 33 | 24.04.2019 | CEO of a for-profit mediator agency | Novac Solutions (joint stock company) |
| 34 | 26.04.2019 | CEO of a for-profit mediator agency | Intermezzo (joint stock company) |
| 35 | 03.05.2019 | City of Zurich, Housing Department | Employee and expert in housing issues, Head of the residential housing section, and expert in non-profit housing provision |
| 36 | 28.05.2019 | City of Kloten, Head of Social Welfare Department | Expert in housing issues and social welfare |
| 37 | 13.06.2019 | Brunaupark's local tenants' association (NGO) | President of the association and resident in Brunaupark |
| 38 | 26.06.2019 | Basel local tenants' association | Head of the association, Expert in housing issues and tenancy law |
| 39 | 08.07.2019 | Credit Suisse pension fund Zurich | Portfolio manager |
| 40 | 31.07.2019 | City of Zurich, Department for Urban Development | Employee and expert in housing issues |
| 41 | 20.08.2019 | Basel-City, Urban Development and Housing Department | Head of the urban development department |
| 42 | 20.08.2019 | City of Köniz, Head of Planning Department | City planner |
| 43 | 12.09.2019 | Credit Suisse asset management Basel | Portfolio manager |

(continued)

(continued)

| *Nr.* | *Date* | *Role* | *Profile of the interviewee* |
|---|---|---|---|
| 44 | 20.09.2019 | Basel-City, Planning Department | City planner |
| 45 | 24.10.2019 | City of Zurich, Head of Planning Department | Head of Planning Department |
| 46 | 18.04.2019 | Resident, Köniz Nessleren | Male, 45 years old, lawyer |
| 47 | 24.04.2019 | Resident, Kloten Southern district | Female, 72 years old, retired |
| 48 | 13.06.2019 | Resident, Köniz Nessleren | Female, 32 years old, housewife |
| 49 | 26.06.2019 | City of Basel, Hirzbrunnen district office | Head of district office |
| 50 | 30.07.2019 | Resident, Kloten southern district | Female, 66 years old, retired |
| 51 | 20.08.2019 | Institutional investor, Previs private foundation | Portfolio manager |
| 52 | 05.09.2019 | Institutional investor, Helvetia insurance | Portfolio manager |
| 53 | 05.09.2019 | Care worker for elderly residents in Basel Schorenweg | Social worker and nurse |
| 54 | 09.09.2019 | Institutional investor, Kloten | CEO of private firm |

# Index

**A**
Active land policy, 63, 64, 177, 195, 196, 200, 202, 262
Actors-centered neoinstitutionalism, 6
Actors use strategies, 34, 35, 118, 128, 248
Added land value capturing, 62, 183
Affordable housing, 3, 4, 30, 61, 62, 64, 69, 91, 98, 120, 131, 158, 163, 166, 171–180, 182–186, 194–204, 207–209, 212, 221, 223, 228, 250, 251, 257, 258, 261, 265
Agenda 21, 2, 47
Analytical framework, 8, 18, 20, 22–25, 34, 43–45, 48, 49, 267, 269
Anti-squat-agencies, 210

**B**
Basel
  Basel Schoren, 95, 99, 107, 152, 153, 155–157, 159, 160, 166, 168, 169
Brundtland report, 49, 50, 150
Bundle of rights, 29
Business of densification, 73, 123, 144, 170, 251, 255, 263, 264

**C**
Case selection, 88, 91, 152, 178, 211, 268
Citizenship, 160
City
  city boundaries, 2, 5, 14, 66, 98, 149, 152, 158, 163, 173
Civil Code, 28, 62, 127, 164, 177, 180, 185, 248
Classic institutionalism, 15
Coherence, 24
Commodification, 31, 152, 171, 178, 203, 207, 209, 210, 217, 264
  commodification of housing, 31, 144, 209, 264
Common law, 23, 28, 177
Common pool resources (CPRs), 16
Commons, 16
Compact city, 1–5, 13, 14, 50, 173, 249, 251, 256, 257, 261, 269



G. Debrunner, *The Business of Densification*,
https://doi.org/10.1007/978-3-031-49014-9

Comparative case study analysis, 119, 152, 177
Complex regimes, 24
Consolidation, 2, 59, 66, 71, 90, 120, 206, 250, 252, 255, 257, 267
Contracts, 28, 62, 64, 67, 98, 106, 111, 135, 151, 155, 160, 164, 168, 173, 176, 177, 184, 185, 199, 204, 206, 210, 220, 221, 225, 226, 230, 249, 251, 257, 265
Cooperative, 28, 32, 62, 64, 69, 132, 144, 186, 197, 219, 252, 265, 269

D
Data collection methods, 103, 110
Decommodification, 68, 69, 210, 263, 264
  decommodification strategies, 67
Densification
  densification goals, 13, 43, 90, 130, 252, 255, 257, 269
  densification objectives, 5, 6, 14, 70, 73, 91, 94, 130, 141, 142, 152, 170, 247, 255
Densifying cities, 70, 172, 198, 203, 205, 264, 269
Displacement, 3, 4, 6, 54, 58, 65, 69, 72, 89, 102, 129, 146, 148, 152, 174, 252, 254, 257

E
Eco-business, 124, 170, 171, 250, 255, 258
Empirical analysis, 8, 88, 99, 180
Engels, Friedrich, 31, 32
Expert interviews, 103, 108, 130, 156
Expropriation, 32, 33, 62, 63, 127, 135, 164, 177, 180, 185, 201
Extension, 4, 140, 226
Extent, 24, 25, 28, 29, 58, 106, 119, 127, 149, 165, 210, 248, 252, 267, 269

F
Federal Constitution, 131
Formal rules, 14, 17, 22, 23, 25, 33, 34, 106, 126, 152, 156, 175, 215, 248, 251, 253, 254, 258, 267, 268

G
Game-actors-play, 34
Gentrification, 3, 4, 6, 68, 69, 89, 125, 129, 130, 151, 163, 209, 213, 256, 257, 267
  new-built gentrification, 91
Global North, 19, 146, 147
Global South, 19
Governance
  governance mechanisms, 8, 44, 70, 87, 91, 94, 103, 117, 119, 120, 247, 261, 262, 267, 268
  governance of densification, 43–45
  governance system, 18
  resource governance, 23
Government, 6, 19, 25, 26, 28, 30, 61, 72, 98, 106, 128, 130, 131, 156, 160, 163–166, 171, 178, 182, 183, 186, 195, 197, 198, 200, 208, 213, 221, 229, 231, 232, 250, 252, 256, 264
Ground leases, 62, 184, 186, 196, 199–201

H
Hardin, Garret, 16
Hobbes, Thomas, 29, 30
Homeowners, 61, 62, 65, 67, 108, 125, 133, 135, 140, 163, 164, 180, 195, 214, 248

Housing
housing affordability, 61, 129, 130, 142, 143, 147, 157, 164, 174, 175, 178, 182–184, 200, 201, 203, 255, 257, 263
housing as a resource, 8, 88, 103, 109, 118, 119, 262
housing availability, 150, 158, 257
housing cooperative, 62, 108, 163, 171, 184, 219, 223, 231, 248, 262, 265
housing policy instruments, 61, 62
housing prices, 4, 105, 131, 135, 149, 153, 213, 226
housing public policies, 60
housing quality, 159
housing stock, 28, 32, 44, 46, 47, 54, 60, 63, 65–67, 69, 71, 89, 96, 102, 106, 119, 125, 135, 136, 140, 147, 149, 152, 161, 163–167, 171–175, 178, 182, 199, 203, 209, 212, 213, 219, 228, 247–249, 253, 256, 263, 267, 268
housing sustainability, 47, 48
housing vacancies, 232

I
Implementation process, 43, 44, 70, 127, 248, 255
Inclusive city, 266, 270
Inequality, 44, 68, 69, 210
Informality, 25
Informal rules, 24, 25, 126
Institutional Analysis and Development (IAD) framework, 18, 22, 23
Institutional economics, 22
Institutionalism, 123, 125, 126, 204, 206
Institutional Resource Regime (IRR) framework, 18, 21–25, 27, 34, 35, 43, 44, 46, 48, 50, 60, 62, 63, 70, 126, 261, 266, 267, 269
Institutions, 8, 14, 15, 17, 18, 20, 23–25, 28, 29, 33, 34, 44, 88, 103, 106, 109, 119, 126, 151, 175, 198, 202, 203, 206, 213, 221, 262, 264
Intensification, 2, 206

J
Joint use situations, 13, 22

K
Kloten
Kloten Southern district, 95, 103–105
Köniz
Köniz Nessleren, 95, 100–102

L
Land
landlord, 5, 31, 135, 136, 145, 220, 248, 251
landowner, 26, 31, 32, 34, 64, 71–73, 91, 94, 119, 120, 127, 141, 142, 144, 151–153, 155, 164, 170, 177, 180, 182–185, 195, 197, 199–201, 203, 204, 206, 209, 211, 215, 219, 220, 229, 249, 253, 255, 258, 265, 269
land policy, 6, 26, 111, 175, 176, 197, 199, 202, 203, 263
land policy instruments, 26, 119, 120, 172, 179, 181, 182, 186, 187, 198, 202
land scarcity, 6, 33, 63, 64, 120, 179, 203, 224, 254, 257, 263, 264

land use planning, 26, 32, 33, 44, 51, 127, 132, 177, 201, 203, 267
land use policy, 26, 60, 65
land use policy instruments, 26
Liberal philosophy, 29
Locke, John, 30
Lower income groups, 4, 96, 99, 102

M
Marx, Karl, 31, 32, 209
Memorial prize, 16
Municipality, 63, 64, 66, 94, 95, 98–100, 102, 103, 105, 153, 161, 176, 182–185, 194, 197, 199, 200, 202, 211, 219, 229, 265
Municipal land policy, 63–65, 72, 181, 200, 202, 253, 254, 263

N
Natural resources, 2, 16, 17, 19–22, 45, 51, 149, 172, 173, 202, 263
Neoinstitutionalism, 15, 16
Neo-Marxism, 20, 21, 48
New institutional economics, 13, 18, 22, 65, 70, 261, 267, 269
New public management, 68
Nobel prize, 16
Non-profit organizations (NGOs), 14, 61, 146, 148, 153, 169, 204, 225, 229
Not-in-my-backyard (NIMBY)
NIMBYism, 69, 252
NIMBY opposition, 69, 72, 252

O
Ostrom, Elinor, 16–18, 22
Our Common Future, 49

P
Participant observation, 103, 106
Passive municipal land policy, 65
Policy goals, 200
Policy instruments, 26, 44, 63, 72, 106, 119, 120, 136, 137, 174, 176–180, 182, 183, 196, 199, 200, 202, 203, 250, 254, 258, 262, 263, 267, 269
Political-administrative actors, 208
Political ecology, 8, 13, 19–21, 43, 70, 111, 261, 267, 269
neo-Marxist political ecology, 20, 21, 48
Political economy, 17, 21
Power, 3, 19–21, 34, 43, 44, 48, 69, 70, 118, 119, 141–145, 152, 176, 197, 201, 207, 208, 211, 220, 229, 231, 248, 249, 253, 254, 257, 262, 265, 268, 269
power games, 13, 14, 44, 70, 111, 247, 261, 267
Pre-emption-rights, 183
Private
private interests, 127, 176, 207
private investors, 66, 67, 153, 169, 174, 183, 184, 194, 196, 197, 202, 252, 265
private law, 27, 28, 61, 62, 65, 73, 131, 136, 161, 165, 176, 180, 184, 186, 196, 201, 203, 207, 213, 217, 219, 254
Property
property owners, 28, 29, 32, 61, 65, 66, 72, 73, 120, 127, 135, 136, 140–144, 157, 160, 165, 169, 178, 182, 198, 203, 206, 207, 213, 214, 217, 218, 220, 221, 223–226, 228, 229, 231, 249, 250, 252, 255, 262
property rights, 13, 15–18, 21–34, 60, 61, 64, 65, 70, 73, 106,

118, 126, 127, 131, 135, 136, 152, 161, 164, 165, 169, 170, 174, 176, 177, 180, 198–203, 206, 207, 213, 217, 220, 248, 251, 254, 258, 261, 262, 267, 269
Public
public interests, 14, 26, 125–127, 135, 164, 168, 182, 185, 207
public ownership, 99, 176, 185, 199, 200
public policy, 13, 25, 26, 33, 61, 66, 70, 127, 130, 176, 177, 180, 198, 201, 203, 213, 248, 261, 262, 267–269
Public laws, 26, 27, 60, 61, 131, 176, 186, 207, 217

**R**
Renovation, 47, 96, 133–135, 173, 182, 225, 248, 256
Rent, 15, 47, 62, 66, 67, 72, 73, 120, 135, 136, 140, 142, 157, 160, 164, 173, 175, 179, 184, 185, 195, 197, 220, 221, 224, 225, 228, 230, 248–252, 255, 256, 258, 265
Rental prices, 6, 96, 98, 212, 257
Replacement construction, 4, 105, 124, 133, 140, 173
Resistance, 21, 66, 68–70, 72, 89, 119, 125, 219, 221, 250, 252, 253, 255, 262
Resource
resource condition, 21, 22, 24, 44, 45, 126, 202
resource users, 21, 33, 70, 262

**S**
Smith, Adam, 30
Social
social exclusion, 3–5, 21, 28, 44, 61, 65, 68, 69, 89, 96, 119, 120, 125, 148, 163, 169, 170, 172, 174, 195, 197, 209, 212, 219, 250, 254, 257, 258, 267
social inclusion, 6, 51, 53, 54, 63, 73, 120, 151, 152, 169, 251, 265
social justice, 51, 54, 126, 129, 206, 262
social sustainability, 3, 6, 49–55, 58, 59, 70, 72, 87, 90, 93, 106, 108, 117–120, 123, 125, 128–130, 136, 142, 143, 145, 147–150, 152, 153, 155–157, 160, 163, 168, 170, 172, 247, 258, 262, 263, 265–269
Socially constructed, 18, 46
Spatial
spatial development, 26, 33, 60, 64, 89, 90, 98, 126, 132, 168, 176, 182, 200, 201
spatial planning, 126, 132, 142
Squatters, 210
Squatting, 207, 208, 213, 215, 218, 224, 229, 252
Stakeholders, 5, 13, 14, 17, 19, 25, 43, 44, 54, 62, 64, 106, 111, 119, 128, 264–269
Strategic
strategic activation, 63, 65, 120, 152, 172, 198, 201–203, 254
strategic planning, 65
Study design, 8, 87, 119
Subsidies, 61, 132, 133, 135, 140, 142, 166, 169, 171, 176, 182, 218, 248, 262, 265
supply-side subsidies, 163, 166, 180, 186, 195, 198
Suburban areas, 4, 96, 153, 158, 159, 171, 179, 202, 212

Sustainability, 19, 21–24, 44, 48–50, 58, 89, 106, 128, 143–145, 149, 150, 171, 256, 257
Sustainable
  sustainable city, 2, 49, 270
  sustainable development, 2, 15, 47–49, 51, 58, 128, 129, 145, 149, 150
Swiss
  Swiss case study, 88
  Swiss federal constitution, 131
  Swiss federal government, 89, 90, 117, 123, 125, 142, 143, 145, 249
  Swiss land use planning, 124, 201
  Swiss tenancy law, 135, 142, 165, 220, 228
Switzerland, 7, 88, 90–92, 94, 95, 98, 99, 118, 124, 125, 129–136, 143, 145–147, 150, 153, 164, 165, 173–175, 178, 184, 185, 204, 210, 211, 226, 248–250, 253, 264, 265, 268

T
Temporary housing, 67, 120, 204–215, 217, 219–221, 223, 224, 226–231, 252, 253, 263
Tenant, 5, 28, 29, 31, 60–62, 67, 69–72, 89, 102, 107, 108, 110, 119, 123, 125, 132, 135, 136, 140–142, 144–149, 152, 153, 155–160, 165, 166, 168, 170, 172–174, 179, 180, 182, 185, 195, 197, 203, 208, 215, 221, 248–253, 257, 258, 268
Theoretical framework, 8, 43

U
United nations (UN)
  UN Agenda 21, 2, 47
  UN Habitat Agenda, 47
Urban
  urban areas, 2, 60, 119, 174, 177, 250, 263
  urban social movements, 212
Urbanization, 1, 88, 89, 171, 253

V
Veto rights, 14, 73, 125, 127, 177, 201, 261, 265

W
Wealth, 4, 20, 89, 130, 169, 178, 197, 210, 254, 256

Z
Zoning, 32, 61, 64, 141, 176, 182, 186, 197–199, 201
  zoning plan, 94, 132, 161, 169, 178, 182–184, 194, 196
Zurich
  Zurich Brunau, 95, 96, 106, 107, 119, 152, 153, 155–162, 164, 166–169

The manufacturer's authorised representative in the EU is Springer Nature Customer Service Centre GmbH, Europaplatz 3, 69115 Heidelberg, Germany. If you have any concerns regarding our products, please contact ProductSafety@springernature.com

Printed and bound by CPI Group (UK) Ltd, Croydon, CR0 4YY

01/10/2026

02225894-0010